Edward Drinker Cope

Check-list of North American Batrachia and Reptilia

with a systematic list of the higher groups, and an essay on geographical distribution. Based on the specimens contained in the U. S. National Museum

Edward Drinker Cope

Check-list of North American Batrachia and Reptilia
with a systematic list of the higher groups, and an essay on geographical distribution. Based on the specimens contained in the U. S. National Museum

ISBN/EAN: 9783337886844

Printed in Europe, USA, Canada, Australia, Japan

Cover: Foto ©Andreas Hilbeck / pixelio.de

More available books at **www.hansebooks.com**

Department of the Interior:
U. S. NATIONAL MUSEUM.

— 1 —

BULLETIN

OF THE

UNITED STATES NATIONAL MUSEUM.

PUBLISHED UNDER THE DIRECTION OF THE SMITHSONIAN INSTITUTION.

WASHINGTON:
GOVERNMENT PRINTING OFFICE.
1875.

North American Batrachia and Reptilia;

WITH A

SYSTEMATIC LIST OF THE HIGHER GROUPS,

AND AN

ESSAY ON GEOGRAPHICAL DISTRIBUTION.

BASED ON

THE SPECIMENS CONTAINED IN THE U. S. NATIONAL MUSEUM.

By EDWARD D. COPE.

WASHINGTON:
GOVERNMENT PRINTING OFFICE.
1875.

ADVERTISEMENT.

This work is the first of a series of papers intended to illustrate the collections of Natural History and Ethnology belonging to the United States and constituting the National Museum, of which the Smithsonian Institution was placed in charge by the act of Congress of August 10, 1846.

It has been prepared at the request of the Institution, and printed by authority of the honorable Secretary of the Interior.

JOSEPH HENRY,
Secretary Smithsonian Institution.

SMITHSONIAN INSTITUTION,
Washington, November, 1875.

TABLE OF CONTENTS.

	Page.
INTRODUCTORY REMARKS	3
PART I. Arrangement of the families and higher divisions of Batrachia and Reptilia. [Adopted provisionally by the Smithsonian Institution.]	7
Class Batrachia	7
Order Anura	7
Stegocephali	10
Gymnophidia	11
Urodela	11
Proteida	12
Trachystomata	12
Class Reptilia	12
Order Ornithosauria	12
Dinosauria	13
Crocodilia	14
Sauropterygia	14
Anomodontia	15
Ichthyopterygia	15
Rhynchocephalia	15
Testudinata	16
Lacertilia	17
Pythonomorpha	
Ophidia	
PART II. Check-list of the species of Batrachia and Reptilia of the Nearctic or North American realm	24
Class Batrachia	24
Order Trachystomata	24
Proteida	24
Caducibranchiata	25
Anura	29
Bufoniformia	29
Firmisternia	30
Arcifera	30
Raniformia	32
Class Reptila	33
Order Ophidia	33
Solenoglypha	33
Proteroglypha	34
Asinea	34
Scolecophidia	44

TABLE OF CONTENTS.

	Page.
PART II. Check-list of the species of Batrachia and Reptilia, &c.—Continued:	
Class Reptila—Continued:	
Order Lacertilia	44
Opheosauri	44
Pleurodonta	44
Typhlophthalmi	44
Leptoglossa	44
Diploglossa	46
Iguania	47
Nyctisaura	50
Testudinata	50
Athecae	50
Cryptodira	51
Crocodilia	54
PART III. On geographical distribution of the Vertebrata of the Regnum Nearcticum, with especial reference to the Batrachia and Reptilia	55
I.—The faunal regions of the earth	55
II.—Number of species	58
III.—Relations to other realms	61
IV.—The regions	67
Austroriparian	68
Eastern	70
Central	71
Pacific	72
Sonoran	73
Lower Californian	74
V.—The Austroriparian region	76
VI.—The Eastern region	82
VII.—The Central region	88
VIII.—The Pacific region	89
IX.—The Sonoran region	90
X.—The Lower Californian region	92
XI.—Relation of distribution to physical causes	93
PART IV. Bibliography	97
A.—Works on the classification of Batrachia and Reptilia	97
B.—Works treating of the geographical distribution of North American Batrachia and Reptilia	100
ALPHABETICAL INDEX	101

INTRODUCTORY REMARKS.

The present contribution to North American Herpetology is a prodromus of a general work on that subject, undertaken some years ago at the request of the Secretary of the Smithsonian Institution. The material which has been accumulating in the museum of that Institution has offered great advantages for the investigation of the questions of anatomical structure, variations of specific characters, and geographical distribution. It is believed that these subjects are much elucidated by the study of the *Batrachia* and *Reptilia*, since these animals are especially susceptible to physical influences; since, also, they are unable, like birds, and generally not disposed, as are mammals, to make extended migrations, their habitats express nearly the simplest relations of life to its surroundings.

In prosecuting these investigations, it has become necessary to adapt the nomenclature to the results obtained by study of many specimens as to the variation of species. It is a common observation that the better a species of animal is represented in our collections, the wider do we discover its range of variation to be, and the greater the number of supposed distinct species does it become necessary to reduce to the rank of varieties. The definition of a species being simply a number of individuals, certain of whose physical peculiarities belong to them alone, and are at the same time exhibited by all of them, it is evident that, since it is impossible, in the present state of our knowledge, to predicate what those "certain peculiarities" shall be, the only test of specific definition is the constancy of those characters. Hence it is that the most diverse forms of one species may differ more from each other than two recognized species. In the investigation of North American cold-blooded *Vertebrata*, I have observed that many species are represented by well-marked geographical varieties, which, following the example of some ornithologists, I have called *subspecies*. Many of these have been heretofore regarded as species.

In illustration of these remarks, certain species of the genus *Ophibolus* may be selected. The most northern and the most southern forms of the genus, the *O. triangulum* and *O. coccineus*, have always been regarded as distinct species; and so numerous are their differential characters, in coloration, size, and squamation, that this view would seem to rest on a satisfactory foundation. I find, however, that individuals exist which represent every stage of development of each character which distinguishes them, although certain types appear to be more abundant than the intermediate ones. *O. triangulum* is a species of larger size, with two temporal plates, a row of large dorsal spots, and other smaller ones on the sides, on a grayish ground ; with a chevron, and often other marks on the top of the head, and a band posterior to the eye. *O. coccineus* is a small snake with a small loreal plate and one temporal shield ; color red, with pairs of black rings extending round the body, and no markings on the head excepting that the anterior ring of the anterior pair crosses the posterior edge of the occipital shields, forming a half collar. The transition is accomplished thus : The lateral borders of the dorsal spots of *O. triangulum* break up, and the lateral spots become attached to their anterior and posterior dark borders. The chevron of the top of the head first breaks into spots, and then its posterior portions unite with each other. The borders of the old dorsal spots continue to the abdomen, where the remaining lateral portions finally meet on the middle line, forming a black line. This breaks up and disappears, leaving the annuli open ; and these are then completed in many specimens. The general colors become more brilliant and the size smaller. The head is more depressed ; in immediate relation to this form, the loreal plate is reduced in size, and the two temporal shields of *O. triangulum* are reduced to one. Every form of combination of these characters can be found, which represent six species of the books (in North America), viz: *O. triangulum, O. doliatus, O. annulatus, O. gentilis, O. amaurus,* and *O. coccineus.* The oldest name is the *O. doliatus,* Linn. Another series of specimens resemble very closely those of the subspecies *coccineus ;* in fact, are identical with them in color. The loreal shield is, however, extinguished, and the rows of scales are reduced by one on each side. These specimens simply carry one degree farther the modifications already described. Yet, on account of the constancy of these characters, I am compelled to regard these individuals not only as a distinct species, but, on account of the absence of the loreal plate, as belonging to another genus. This is the *Calamaria clapsoidea* of Holbrook ; the *Osceola elapsoidea* of Baird and Girard. It affords an illustration of the principle, which I have elsewhere insisted on, " that adjacent species of allied genera may be more alike than remote species of identical generic characters," which indicates that generic characters originate independently of the specific.*

The classification of the present list is illustrated by the above remarks. I now briefly allude to the rules I have followed in adopting a nomenclature. These rules are those in general use in the United States, as based on the revision of the rules of the British Association for the Advancement of Science by a committee of the American Association, and elaborated in more detail by W. H. Edwards,† after Thorell and Wallace; in other words, the law of priority is followed under the following definitions:

(1) A specific name given by an author must relate to a description or plate of the object intended.

(2) A generic name of a species must be accompanied by a separate definition of the genus intended, by reference to some of its distinctive features.

NOTE.—These two rules are properly regarded as the safeguards of nomenclature, since they offer the only means by which the writings of authors in the sciences concerned can be intelligible. The necessity of these rules will become increasingly apparent, since, as the systematic sciences become more popular, sciolists may publish pages of names in any of their departments, with the effect, should such names be authoritative, of indefinitely postponing the cultivation of the subject. A generic diagnosis is not necessarily perfect in the early stages of the classification of a science, and may be found later to embrace more than one generic type; hence, the following additional rule has been found necessary:

(3) In the subdivision of a genus, names of the new genera are to be adopted in the order of priority of the definition of the divisions to which they refer; the remaining natural generic group retaining the original name, unless the latter has been already given to one of the divisions, as prescribed.

(4) Priority reposes on date of publication, and not on date of reading of papers.

Of course, consistently with the above rules, as divisions of high rank must be defined in order to be understood, names of these unaccompanied by definitions are not binding on the nomenclator.

In regard to orthography, the same code of rules has been followed, viz, in the Latinization of all words of Greek derivation. This has been

* Origin of Genera, Philadelphia, 1868.
† The Canadian Entomologist, 1873, p. 32.

applied especially to the compounding of family-names. Thus, if the generic name is spelled according to Latin rule, the family-name derived from it must be so also; hence, I write *Scaphiopidae*, not Scaphiopodidae; *Rhinoceridae*, not Rhinocerotidae.

In the check-list, the correct name of each species and subspecies is given with reference to a good description. To each is added its geographical range.

PART I.

ARRANGEMENT

OF

THE FAMILIES AND HIGHER DIVISIONS

OF

BATRACHIA AND REPTILIA.

[ADOPTED PROVISIONALLY BY THE SMITHSONIAN INSTITUTION.]

CLASS BATRACHIA.

Order ANURA.

(Anura, Duméril; Salientia, Merrem, Gray.)

RANIFORMIA.

(Raniformia, Cope, Nat. Hist. Rev., v, 114, 1865.[1])

Ranidae	= Ranidae, Cope, N. H. Rev., v, 114–119, 1865.[2]
Colostethidae	= Colostethidae, Cope, P. A. N. S. Phila., 1866, 130.[3]

---

[1] Raniformia, partim, Dum. et Bib., Erp. Gén.

[2] Ranidae, Cope, Jour. Acad. Nat. Sci. Phila., n. s., vi, 189, 1867; Ranidae, Polypedatidae, and Cystignathidae, pars, Gthr., Cat. Bat. Salien., 1858, 4–26.

[3] Colostethidae, Cope, Jour. Acad. Nat. Sci. Phila., n. s., vi, 197, 1867; "Calostethidae," Mivart, Proc. Zoöl. Soc. London, 1869.

FIRMISTERNIA.[4]

(Bufonoid Raniformia, Cope, Jour. Acad. Nat. Sc. Phila., n. s., vi, 190, 1867.)

Dendrobatidae	= Dendrobatidae, Cope, N. H. Rev., v, 103–104, 1865.[5]
Phryniscidae	= Phryniscidae, Cope, J. A. N. S. Phila., n. s., vi, 190, 1867.[6]
Engystomidae	= Engystomidae, Cope, J. A. N. S. Phila., n. s., vi, 190, 1867.[7]
Brevicipitidae	= Brevicipitidae, Cope, J. A. N. S. Phila., n. s., vi, 190, 1867.[8]

GASTRECHMIA.

(Gastrechmia, Cope, J. A. N. S. Phila., n. s., vi, 198, 1867.)

Hemisidae	= Hemisidae, Cope, J. A. N. S. Phila., n. s., vi, 198–199, 1867.[9]

---

[4] Firmisternia. Believing the arciferous or raniform sternal structure to have about equal systematic value with the presence or absence of teeth, I have separated the toothless families with raniform sternum under the name of Firmisternia. It is not impossible that this group may turn out to be inseparable from the Gastrechmia. The toothed Aglossa must be distinguished on the same principle from Pipa, and the suborder is accordingly named Odontaglossa.

[5] Hylaplesiidae, Gthr., Cat. Bat. Salien., 1858, 124–126.

[6] Brachycephalina, pars, Gthr., Cat. Bat. Salien., 1858, 42.

[7] Engystomidae, Cope, N. H. Rev., v. 100–101, 1865; Michrylidae, Brachymeridae, Eugystomatidae, Hylaedactylidae, Gthr., Cat. Bat. Salien., 1858.

[8] Brachymeridae, Cope, pars, N. H. Rev., v, 101–102, 1865.

[9] Hemisidae; Rhinophrynidae, Cope, pars, N. H. Rev., v, 100, 1865; Rhinophrynidae et Phryniscidae, pars, Mivart, Proc. Zoöl. Soc. London, 1869, 281–222.

BUFONIFORMIA.

(Bufoniformia, Duméril et Bibron, partim; Cope, partim.)

Rhinophrynidae = Rhinophrynidae, Gthr., Cat. Bat. Sal. B. M., 127, 1858.[10]

Bufonidae = Bufonidae, Cope, N. H. Rev., v, 102–103, 1865.[11]

Batrachophrynidæ = Batrachophrynus, Peters, Monatsb. Pr. Akad. Wiss., 1873, 411.

AGLOSSA.

Pipidae = Pipidae, Gthr., Cat. Bat. Sal. B. M., 2–3, 1858.[12]

ODONTAGLOSSA.

Dactylethridae = Dactylethridae, Gthr., Cat. Bat. Sal. B. M., 1–2, 1858.[13]

ARCIFERA.

(Arcifera, Cope, N. H. Rev., v, 104, 1865.[14])

Cystignathidae = Cystignathidae, Cope, N. H. Rev., v, 105, 1865.[15]

---

[10] Rhinophrynidae, Cope, N. H. Rev., v, 100, 1865, pars, nec Mivart; Cope, Jour. Acad. Nat. Sci. Phila., vi, 189, 1867.

[11] (Bufonidae) Chelydobatrachus, Gthr., Cat. Bat. Salien., part., 1858, 51, 53–54.

[12] Pipidae, Cope, N. H. Rev., v, 98–99, 1865; Pipidae, Mivart, Proc. Zoöl. Soc. London, 1869, 287, 295.

[13] Dactylethridae, Cope, N. H. Rev., v, 99, 1865; Dactylethridae, Mivart, Proc. Zoöl. Soc. London, 1869, 295.

[14] Arcifera, Cope, Jour. Nat. Sci. Phila., vi, 67–68, 1866.

[15] Cystignathidae, Ranidae partim, Cystignathidae, Uperoliidae, Bombinatoridae partim, Alytidae partim, Hylodidae, Gthr.; Ranidae partim, Polypedatidae partim, Discoglossidae partim, Mivart, Proc. Zoöl. Soc. London, 1869.

Hemiphractidae = Hemiphractidae, Cope, J. A. N. S. Phila., n. s., vi, 69, 1866.
Hylidae > Hylidae, Gthr., Cat. Bat. Salien., 96, 1858.[16]
Scaphiopidae = Scaphiopodidae, Cope, J. A. N. S. Phila., n. s., vi, 69, 1866.[17]
Pelodytidae = Pelodytidae, Cope, J. A. N. S. Phila., vi, 69, 1866.[18]
Asterophrydidae = Asterophrydidae, Cope, J. A. N. S. Phila., n. s., vi, 79-80.[16a]
Discoglossidae = Discoglossidae, Cope, N. H. Rev., v, 105-107, 1865.[19]

Order STEGOCEPHALI.

(Stegocephali, Cope, P. A. N. S. Phila., 1868, 209.[20])

LABYRINTHODONTIA.

Baphetidae = Baphetidae, Cope, MSS.
Anthracosauridae = Anthracosauridae, Cope, MSS.

GANOCEPHALA.

Colosteidae = Colosteidae, Cope, MSS.[21]

[16] Hylidae, Cope, T. A. N. S. Phila., vi, 83-85, 1866.
[17] Scaphiopodidae partim, N. H. Rev., v, 107-108, 1865.
[18] Pelodytidae. Scaphiopodidae pars, Cope, olim, Jour. Acad. Nat. Sci. Phila., vi, 69, 1866.
[19] Discoglossidae, Cope, Jour. Acad. Nat. Sci. Phila., vi, 69, 1866; Discoglossidae partim, 34, Bombinatoridae partim et Alytidae partim Gthr., Cat. Bat. Salien., 40, 57, 1858; Mivart, Proc. Zoöl. Soc. London, 1869, 291-295.
[20] Stegocephali, Cope, Trans. Am. Phil. Soc. 1870, 6-7.
[21] Colosteus, Cope.

MICROSAURIA.

Phlegethontiidae	= *Phlégethontiidae*, Cope, MSS.[21a]
Molgophidae	= *Molgophidae*, Cope, MSS.[22]
Ptyoniidae	= *Ptyoniidae*, Cope, MSS.[23]
Tuditanidae	= *Tuditanidae*, Cope, MSS.
Peliontidae	= *Peliontidae*, Cope, MSS.[24]

Order GYMNOPHIDIA.

(Gymnophiona, Müller.)

Caeciliidae	= Caeciliidae, Gray, Cat. Bat. Grad. B. M., 57, 1850.

Order URODELA.

Pleurodelidae	= { Seiranotidae, Pleurodelidae, } Gray, P. Z. S. London, xxvi, 137–143, 1858.
Salamandridae[25]	= Salamandridae, Gray, P. Z. S. London, xxvi, 142–143, 1858.
Hynobiidae[26]	= Hynobiidae, Cope, J. A. N. S. Phila., n. s., vi, 107, 1866.
Desmognathidae	= Desmognathidae, Cope, J. A. N. S. Phila., n. s., vi, 107, 1866.
Thoriidae	= Thoriidae, Cope, P. A. N. S. Phila., 1869, 111–112.

[21a] Phlegethontia, Cope.
[22] Molgophis, Cope.
[23] Lepterpeton, Huxl.; Oestocephalus, Cope; Urocordylus, Huxl.
[24] Pelion, Wyman.
[25] Salamandridae, Cope, Jour. Acad. Nat. Sci. Phila., vi, 107–108, 1866.
[26] Hynobiidae, Cope; Molgidae, Gray, 1850.

Plethodontidae[27] = Plethodontidae, Cope, J. A. N. S. Phila., n. s., vi, 106-107, 1866.
Amblystomidae[28] = Amblystomidae, Cope, J. A. N. S. Phila., n. s., vi, 105-106, 1866.
Menopomidae = Protonopsidae, Gray, Cat. Bat. Grad. B. M., 52-54, 1850.
Amphiumidae = Amphiumidae, Cope, J. A. N. S. Phila., n s., vi, 104-105, 1866.
Cocytinidae = Cocytinidae, Cope, MSS.[29]

Order PROTEIDA.

Proteidae = Proteidae. Gray, Cat. Bat. Grad. B. M., 64-67, 1850.

Order TRACHYSTOMATA.

Sirenidae = Sirenidae. Gray, Cat. Bat. Grad. B. M., 67-69, 1850.

Class REPTILIA.

Order ORNITHOSAURIA.

(Ornithosauria, Bonaparte, Fitzinger, Seeley.[30])

Dimorphodontidae = Dimorphodontidae, Cope, P. A. A. A. S. 1870, 234, 1871.[31]

[27] Plethodontidae, Cope, Jour. Acad. Nat. Sci. Phila., vi, 106, 1866, partim Gray, 1850.
[28] Amblystomidae. Plethodontidae partim, Gray, 1850.
[29] Cocytinus, Cope, Trans. Am. Philos. Soc. Phila., 1874.
[30] Ornithosauria = Pterosauria, Owen.
[31] Dimorphodontae, Seeley.

Pterodactylidae = Pterodactylidae, Cope, P. A. A. A. S., xix, 234, 1871.[32]

Order DINOSAURIA.

(Dinosauria, Owen, Cope, Seeley; *Pachypodes*, Meyer; Ornithoscelida, Huxley.)

SYMPHYPODA.

(Symphypoda, Cope; Compsognatha, Huxley.)

Compsognathidae = Compsognathidae, Cope, P. A. A. A. S., xix, 234, 1871[33] (name only).

Ornithotarsidae = Ornithotarsidae, Cope, P. A. A. A. S., 234, 1871[34] (name only).

GONIOPODA.

(Goniopoda, Cope; Harpagmosauria, Haeckel.)

Megalosauridae = Megalosauridae, Cope, P. A. A. A. S., xix, 234, 1871 (name only).[35]

Teratosauridae = Teratosauridae, Cope, P. A. A. A. S., xix, 234, 1871 (name only).[36]

ORTHOPODA.

(Orthopoda, Cope; Therosauria, Haeckel.)

Scelidosauridae = Scelidosauridae, Cope, T. A. P. S., n. s., xiv, 91, 1869.[37]

---

[32] Rhamphorhynchae et Pterodactylae, Seeley, loc. cit.
[33] Compsognathidae = Compsognathus, Wag.
[34] Ornithotarsidae = Ornithotarsus, Cope.
[35] Megalosauridae, Huxley.
[36] Teratosaurus, Plateosaurus, Meyer, etc.
[37] Scelidosauridae, Huxley, Journ. Geol. Soc. London, 1870.

Iguanodontidae	= *Iguanodontidae*, Cope, T. A. P. S., n. s., xiv, 91, 1869.[32]
Hadrosauridae	= *Hadrosauridae*, Cope, T. A. P. S., n. s., xiv, 91–98, 1869.[33]

Order CROCODILIA.

(Crocodilia et Thecodontia, partim, Owen, 1841.)

PARASUCHIA.

Belodontidae	= *Belodontidae*, Cope, P. A. A. A. S., xix, 234, 1871 (name only).[40]

AMPHICOELIA.

Teleosauridae	= *Teleosauridae*, Cope, P. A. A. A. S., xix, 234, 1871 (name only).
Goniopholididae	= *Goniopholis*, Owen, etc.

PROCOELIA.

Thoracosauridae	= *Thoracosauridae*, Cope, P. A. A. A. S., xix, 235, 1871 (name only).[41]
Crocodilidae	= Crocodilidae, Cope, P. A. A. A. S., xix, 235, 1871 (name only).[42]

Order SAUROPTERYGIA.

(Sauropterygia, Owen.)

? *Placodontidae*	= *Placodontidae*, Cope, P. A. A. A. S., xix, 235, 1871 (name only).[43]

[32] Iguanodontidae, Huxley, Journ. Geol. Soc. London, 1870.
[33] Hadrosauridae, Huxley, Journ. Geol. Soc. London, 1870.
[40] Thecodontia, Owen, pt.; Cope, Tr. A. P. S., 1869, 32.
[4] Thoracosaurus, Leidy, Cope.
[42] Crocodilidae + Alligatoridae, Gray, + Gavialidae, Gray, + Holops and Thecachampsa, Cope, etc., Pr. A. A. A. S., xix, 235, 1871.
[43] Placodus, Agass.

Plesiosauridae	= *Plesiosauridae*, Cope, P. A. A. A. S., xix, 235, 1871 (name only).[44]
Elasmosauridae	= *Elasmosauridae*, Cope, Tr. A. P. S., n. s., xiv, 1869, p. 47.[45]

Order ANOMODONTIA.

(Anomodontia, Owen.)

Dicynodontidae	= *Dicynodontidae*, Cope, P. A. A. A. S., xix, 235, 1871 (name only).[46]
Oudenodontidae	= *Oudenodontidae*, Cope, P. A. A. A. S., xix, 235, 1871 (name only).[47]

Order ICHTHYOPTERYGIA.

Ichthyosauridae	= *Ichthyosauridae*, Cope, P. A. A. A. S., xix, 235, 1871.

Order RHYNCHOCEPHALIA.

Protorosauridae	= *Protorosauridae*, Cope, P. A. A. A. S., xix, 235, 1871 (name only).[48]
Sphenodontidae	= Sphenodontidae, Cope, P. A. A. A. S., xix, 235, 1871.[49]
Rhynchosauridae	= *Rhynchosauridae*, Cope, P. A. A. A. S., xix, 235, 1870 (name only).[50]

[41] Nothosaurus, Pistosaurus, Plesiosaurus, Pliosaurus, etc.
[45] Elasmosaurus, Cimoliasaurus, etc.
[46] Dicynodontidae, Owen, Palæontology.
[47] Cyptodontia, Owen, Palæontology.
[48] Protorosaurus, Meyer (elongate sacrum).
[49] Hatteriidae, Cope, Proc. Acad. Nat. Sc. Phila., 1864, 225-7.
[50] Rhynchosaurus, Owen.

Order TESTUDINATA.

ATHECAE.

(Athecae, Cope. P. A. A. A. S., xix, p. 235, 1870.)

Sphargididae	= Sphargididae, Gray, Ann. Philos., 1825.[51]
Protostegidae	= Protostega, Cope, Proc. A. P. S., 1872, 413.

CRYPTODIRA.

Cheloniidae	= Cheloniidae, Gray, Annals Philosophy, 1825.[52]
Propleuridae	= Propleuridae, Cope, Am. Jour. Sc. and Arts, I. 137, 1870.
Trionychidae	= Trionychidae, Gray, Annals of Philosophy, 1825.[53]
Emydidae	= Emydidae, Agassiz, Cont. Nat. Hist. U. S., i, p. 351.[54]
Chelydridae	= Chelydridae, Agassiz, Contrib. N. H. U. S., i, 341.[54a]
Cinosternidae	= Cinosternidae, Agassiz, Cont. Nat. Hist. U. S., i, 347.
Testudinidae	= Testudinidae, Cope, P. A. N. S. Phil., 1868, p. 282.[55]

[51] Sphargididae, Bell, Fitzinger, Agassiz.
[52] Cheloniidae, Gray, Ann. Phil., 1825; Agass., Cope, P. A. A. A. S., xix, 235, 1871.
[53] Trionychidae, Bell, Wiegmann, Dum. et Bibr., Agass.
[54] Emydidae—Chelydridae, Cope, P. A. A. A. S., xix, 235, 1871 (name only).
[54a] Chelydra, Cope, P. A. N. S. Phila., 1872.
[55] Testudinidae, Gray, Agass.

Pleurosternidae	= *Pleurosternidae*, Cope, P. A. N. S. Phila., 1868, 282 (name only).
Adocidae	= *Adocidae*, Cope, P. A. P. S., 1870, 547.

Pleurodira.

(Pleurodira, Dum. et Bibron; Chelyoidae, Agass.)

Podocnemididae	= Podocnemididae, Cope, P. A. N. S. Phila., 1868, 282.
Chelydidae	= Chelydidae, Gray, P. Z. S. London, 1869, pp. 208–209.
Hydraspididae	= Hydraspididae, Cope, P. A. N. S. Phila., 1868, 282.
Pelomedusidae	= Pelomedusidae, Cope, P. A. N. S. Phila., 1865, 185; 1868, p. 119.
Sternothaeridae	= Sternothaeridae, Cope, P. A. N. S. Phila., 1868, 119.

Order LACERTILIA.

(Lacertilia, Owen; Cope, P. A. A. A. S., xix, 236, 1870.)

Rhiptoglossa.

(Acrodonta Rhiptoglossa, Wiegmann, Fitzinger, Cope; Chamaeleonida, Müller.)

Chamaeleontidae	= Chamaeleontidae, Gray, Cat. Lizards B. M., 1845, 264 (name only).[56]

---

[56] Wiegmann, Gray, etc.

Pachyglossa.

(Pachyglossa, Cope; Acrodonta Pachyglossa, Wagler, Fitzinger, Cope, P. A. N. S. Phila., 1864, 226-227.)

Agamidae = Agamidae, Gray, Cat. B. M., 1845, 230.

Nyctisaura.

(Nyctisaura, Gray, Cat. Lizards B. M.; Cope, P. A. N. S. Phila., 1864, 225.)

Gecconidae = Gecconidae, Gray, Cat. Lizards B. M., 1845, 142.[57]

Pleurodonta.

(Pleurodonta, Cope, P. A. N. S. Phila., 1864, 226.)

a. *Iguania.*

Anolidae = Anolidae, Cope, P. A. N. S. Phila., 1864, 227, 228.
Iguanidae = Iguanidae, Cope, P. A. N. S. Phila., 1864, 227, 228.[58]

b. *Diploglossa.*

Anguidae = Anguidae, Cope, P. A. N. S. Phila., 1864, 228.
Gerrhonotidae = Gerrhonotidae, Cope, P. A. N. S. Phila., 1864, 228.[59]

---

[57] Cope, Pr. A. A. A. S., xix, 236, 1871.
[58] Iguanidae pars auctorum.
[59] Zonuridae, pt., Gray.

Xenosauridae = Xenosauridae, Cope, P. A. N. S. Phila., 1866, 322.
Helodermidae = Helodermidae, Gray, Cat. Lizards B. M., 1845.[60]

c. *Thecaglossa.*

(Thecaglossa, Wagler, Fitzinger, Cope.)

Varanidae = Varanidae, Cope, P. A. A A. S., xix, 237, 1870.

d. *Leptoglossa.*

(Leptoglossa, Wiegmann, Fitzinger, Cope.)

Teidae = Teidae, Cope, P. A. A. A. S., xix, 237, 1871.[61]
Lacertidae = Lacertinidae, Gray, Cat. Lizards B. M., 26–44, 1845.[62]
Zonuridae = Zonuridae, Cope, P. A. A. A. S., xix, 237–241, 1871.[63]
Chalcidae = Chalcidae, Gray, Cat. Lizards B. M., 57–58, 1845.[64]
Scincidae = Scincidae, Gray, Cat. Lizards B. M., 70–120, 1845.[65]
Sepsidae = Sepsidae, Gray, Cat. Lizards B. M., 121–126, 1845.[66]

---

[60] Helodermidae, Cope, Proc. Acad. Nat. Sc. Phila., 1864, 228; 1866, 322.
[61] Teidae and Ecpleopodidae, Peters, Cope (Proc. Acad. Nat. Sci. Phila., 1864, 229); Teidae, Anadiidae, Cercosauridae, Riamidae, Gray.
[62] Lacertidae, Cope, Proc. Acad. Nat. Sci. Phila., 1864, 228; Lacertidae et Cricosauridae, Peters; Xantusiidae, Baird.
[63] Zonuridae, pt., Gray; Lacertidae pt., Cope.
[64] Chalcididae, Cope, Proc. Acad. Nat. Sci. Phila., 1864, 228.
[65] Scincidae, Cope, Proc. Acad. Nat. Sci. Phila., 1864, 228.
[66] Sepsidae, Cope, Proc. Acad. Nat. Sci. Phila., 1864, 228.

c. *Typhlophthalmi.*

(Typhlophthalmi, Cope, P. A. N. S. Phila., 1864, 228.[67])

Feyliniidae	= Anelytropidae, Cope, P. A. N. S. Phila., 1864, 230.[68]
Acontiidae	= Acontiadae, Gray, Cat. Lizards B. M., 126–127, 1845.[69]
Aniellidae	= Aniellidae, Cope, P. A. N. S. Phila., 1864, 230.

OPHEOSAURI.

(Opheosauri, Cope, P. A. N. S. Phila., 1864, 226.[70])

Amphisbaenidae	= Amphisbaenidae, Gray, Cat. Tort. Croc., etc. B. M., 69, 1844.[71]
Trogonophidae	= Trigonophidae, Gray, Catal. Tort. Croc., etc. B. M., 68, 1844.[72]

Order PYTHONOMORPHA.

(Pythonomorpha, Cope, T. A. P. S., n. s., xiv, 175–182, 1870.[73])

Mosasauridae	> *Mosasauridae*, Cope, T. A. P. S., n. s., xiv, 182–211, 1870.

---

[67] Typhlophthalmi, pars., Dum. et Bib., Erp. Gen.
[68] Typhlinidae, Gray.
[69] Acontiidae, Cope, Proc. Acad. Nat. Sci. Phila., 1864, 230.
[70] Ophisauri, Merrem; Annulati, Wiegmann; Ptychopleures Glyptodermes, Dum. et Bib.; Amphisbaenoidea, Müller.
[71] Amphisbaenidae, Wiegmann.
[72] Trogonophes, Wiegmann, Fitzinger.
[73] Pythonomorpha, Cope, Proc. Bost. Nat. Hist. Soc., 1869, 251; **Lacertilia Natantia**, Owen, Paleontographical Soc. Cretaceous Reptiles.

Order OPHIDIA.

Scolecophidia.

(Scolecophidia, Dum. et Bib.[74])

Typhlopidae	= Typhlopidae, Cope, P. A. A. A. S., xix, 237, 1871 (name only).[75]
Stenostomidae	= Stenostomidae, Cope, P. A. A. A. S., xix, 237, 1871 (name only).[76]

Tortricina.

(Tortricina, Müller.[77])

Tortricidae	= Tortricidae, Cope, P. A. N. S. Phila., 1864, 230.
Uropeltidae	= Uropeltidae, Cope, P. A. N. S. Phila., 1864, 230.[78]

Asinea.

(Asinea, Müller, Cope.)

a. *Peropoda.*

(Peropoda, Müller.)

Xenopeltidae	= Xenopeltidae, Cope, P. A. N. S. Phila., 1864, 230.[79]
Pythonidae	= Pythonidae, Cope, P. A. N. S. Phila., 1864, 230.[80]

---

[74] Scolecophidia et Catodonta, Cope, Proc. Acad. Nat. Sci. Phila., 1864, 230.
[75] Epanodontiens, Dum. et Bib.
[76] Catodontiens, Dum. et Bib.; Catodonta, Cope, olim.
[77] Tortricina, Cope, Proc. Acad. Nat. Sci. Phila., 1864, 230.
[78] Uropeltacea, Peters; Rhinophidae, Gray.
[79] Xenopeltidae, Gthr., Reptiles British India.
[80] Holodontiens, Dum. et Bib.

Boidae = Boidae, Cope, P. A. N. S. Phila., 1864, 230.[1]

Lichanuridae = Lichanuridae, Cope, P. A. N. S. Phila., 1868, 2.

b. *Colubroidea.*

Achrochordidae = Achrochordidae, Cope, P. A. N. S. Phila., 1864, 231.[2]

Homalopsidae = Homalopsinae, Cope, P. A. N. S. Phila., 1864, 167.[3]

Colubridae = Colubridae, Cope, P. A. A. A. S., xix, 238, 1870.[4]

Rhabdosomidae = Rhabdosomidae, Cope, P. A. A. A. S., xix, 238, 1870.[5]

PROTEROGLYPHA.

a. *Conocerca.*

Elapidae = Elapidae, Cope, P. A. N. S. Phila., 1864, 231.[6]

Najidae = Najidae, Cope, P. A. N. S. Phila., 1864, 231.[7]

[1] Aproterodontiens, Dum. et Bib.
[2] Achrochordiens, Dum. et Bib.
[3] Natricidae, pars, Gthr., Cat. Col. Snakes B. M., 1858, 50–84, Potamophilidae, Jan.
[4] Asinea, Group β-bb, Cope, Proc. Acad. Nat. Sci. Phila., 1864, 231; Calamaridae, Olgodontidae, Coronellidae, Colubridae, Dryadidae, Dendrophididae, Dryiophididae, Psammophididae, Lycodontidae, Scytalidae, Dipsadidae, etc., Gthr., Cat. Col. Snakes B. M., 1858, et op. alt.
[5] Calamaridae partim, Gthr., Cat. Col. Snakes B. M., 1858, 2-22.
[6] Elapidae (pars), Gthr., Cat. Col. Snakes B. M., 1858, 209-237.
[7] Elapidae (pars altera), Gthr., Cat. Col. Snakes B. M., 1858, 209-237.

b. *Platycerca.*

Hydrophidae	= Hydridae, Gray, Cat. Snakes B. M., 2, 35, 40, 1849.[88]

SOLENOGLYPHA.[89]

(Solenoglypha, Dum. et Bib.)

Atractaspididae	= Atractaspididae, Gthr., Cat. Snakes B. M., 239, 1858.[90]
Causidae	= Causidae, Cope, P. A. N. S., Phila., 1859, 334.
Viperidae	= Viperidae, Gray, Cat. Brit. Mus., p. 18.[91]
Crotalidae	= Crotalidae, Gray, Cat. Brit. Mus.[92]

---

[88] Hydridae, Gray; Hydrophidae, Schmidt, Fischer; Cope, Proc. Acad. Phila., 1859, 333.

[89] Viperidae, Cope, Proc. Acad. Nat. Sci. Phila., 1859, 333.

[90] Atractaspidinae, Cope, Proc. Acad. Nat. Sci. Phila., 1859, 334.

[91] Viperinae, Cope, Proc. Acad. Nat. Sci. Phila., 1859; Günther.

[92] Crotalinae, Cope, Proc. Acad. Nat. Sci. Phila., 1859; Günther, Cat. Col. Snakes B. M. et auctorum.

PART II.

CHECK-LIST

OF

THE SPECIES OF BATRACHIA AND REPTILIA

OF

THE NEARCTIC OR NORTH AMERICAN REALM.

BATRACHIA.

TRACHYSTOMATA.

SIRENIDAE.

SIREN, Linn.

Siren lacertina, Linn.; Holbrook, N. Am. Herpetology, vol. v, p. 101. The Austroriparian region; extreme points North Carolina, Florida, Matamoras, Mexico, and Alton, Illinois.

PSEUDOBRANCHUS, Gray.

Pseudobranchus striatus, LeConte: Holbrook, American Herpetology, vol. v, p. 109. Georgia.

PROTEIDA.

PROTEIDAE.

NECTURUS, Raf.

Necturus lateralis, Say; Holbrook, Am. Herp., vol. v, pp. 111, 115. Eastern region except New England and eastern Middle States; from a few points in the Austroriparian.

Necturus punctatus, Gibbes. Eastern South Carolina.

CADUCIBRANCHIATA.

AMPHIUMIDAE.

AMPHIUMA, Linn.

Amphiuma means, Linn.; Holbrook, Am. Herp., v, p. 89. Austroriparian region, from North Carolina to Mississippi.

MURAENOPSIS, Fitzinger.

Muraenopsis tridactylus, Cuvier; Holbrook, Am. Herp., v, p. 93. Mississippi and Louisiana.

MENOPOMIDAE.

MENOPOMA, Harl.

Menopoma alleghaniense, Harl.; Holbrook, Am. Herp., v, p. 95. All tributaries of the Mississippi, and streams of the Louisianian district to North Carolina.

Menopoma fuscum, Holbrook, Am. Herp., v, p. 99. Headwaters of the Tennessee River.

AMBLYSTOMIDAE.

AMBLYSTOMA, Tschudi.

Amblystoma talpoideum, Holbrook; Cope, Proceedings Academy Philadelphia, 1867, p. 172. Austroriparian region; mountains of South Carolina.

Amblystoma opacum, Gravenhorst; Cope, Proceed. Acad. Phila., 1867, p. 173. From Pennsylvania to Florida, to Wisconsin, and to Texas.

Amblystoma punctatum, Linn.; Cope, loc. cit., 1867, p. 175. United States, east of the plains; Nova Scotia.

Amblystoma conspersum, Cope, loc. cit., 1867, 177. Pennsylvania to Georgia.

Amblystoma bicolor, Hallowell; Cope, loc. cit., 178. New Jersey.

Amblystoma tigrinum, Green; Cope, loc. cit., 179. United States, east of the plains.

Amblystoma mavortium, Baird; Cope, loc. cit., 184. United States, in the Central, Sonoran, and Pacific regions.

Amblystoma mavortium, Baird; subspecies *californiense*, Gray; Cope, loc. cit., p. 187. Pacific region.

Amblystoma obscurum, Baird; Cope, loc. cit., p. 192. Iowa.

Amblystoma xiphias, Cope, loc. cit., p. 192. Ohio.

Amblystoma trisruptum, Cope, loc. cit., p. 194. New Mexico.

Amblystoma jeffersonianum, Green, subspecies *jeffersonianum*, Green; Cope, loc. cit., p. 195. Pennsylvania and Ohio, and northward.

Amblystoma jeffersonianum, Green, subspecies *laterale*, Hallowell; Cope, loc. cit., p. 197. Canada and Wisconsin, and northward.

Amblystoma jeffersonianum, Green, subspecies *fuscum*, Hallowell; Cope, loc. cit., 197. Indiana and Virginia.

Amblystoma jeffersonianum, Green, subspecies *platineum*; Cope, loc. cit., p. 198. Ohio.

Amblystoma macrodactylum, Baird; Cope, loc. cit., p. 198. Pacific region.

Amblystoma paroticum, Baird; Cope, loc. cit., p. 200. Vancouver's Island and Washington Territory.

Amblystoma aterrimum, Cope, loc. cit., p. 201. Northern Rocky Mountains.

Amblystoma tenebrosum, Baird and Girard; Cope, loc. cit., p. 202. Pacific region of Oregon and California.

Amblystoma texanum, Matthes; Cope, loc. cit., p. 204. Texas.

Amblystoma cingulatum, Cope, loc. cit., p. 205. South Carolina.

Amblystoma microstomum, Cope, loc. cit., p. 206. Austroriparian and Eastern regions, west of the Alleghany Mountains.

DICAMPTODON, Strauch.

Dicamptodon ensatus, Eschscholz, Zoölogical Atlas, part v, p. 6, pl. xxii. Pacific region.

PLETHODONTIDAE.

BATRACHOSEPS, Bonap.

Batrachoseps attenuatus, Eschscholz, Hallowell, Jour. Acad. Phila., 1858, p. 348. Pacific region.

Batrachoseps nigriventris, Cope, Proceed. Acad. Phila., 1869, p. 98. Fort Tejon, California.

Batrachoseps pacificus, Cope, Proceed. Acad. 1865, p. 195. Santa Barbara, Cal.

HEMIDACTYLIUM, Tschudi.

Hemidactylium scutatum, Schlegel; Duméril et Bibron, Erp. Générale, ix, p. 118–9. Rhode Island to Illinois, and to the Gulf of Mexico.

PLETHODON, Tschudi.

Plethodon cinereus, Green, subspecies *cinereus*, Green; Cope, Proceed. Acad. Phila., 1869, p. 99. Eastern region.

Plethodon cinereus, Green, subspecies *erythronotus*, Green; Holbrook, N. Am. Herp., v, p. 43. Eastern region.

Plethodon cinereus, Green, subspecies *dorsalis*, Baird, MSS. Louisville, Ky.; Salem, Mass.

Plethodon intermedius, Baird, Proceed. Acad. Phila., 1857, p. 209. Vancouver's Island.

Plethodon glutinosus, Green; Cope, loc. cit., 1869, p. 99. Eastern and Austroriparian regions.

Plethodon oregonensis, Girard; Cope, loc. cit., p. 99. Pacific region.

Plethodon flavipunctatus, Strauch., Mem. Acad. Sci. St. Petersburg, 1871, xvi, 71. ? New Albion, Cal.

Plethodon croceater, Cope, loc. cit., 1857, p. 210. Lower California.

STEREOCHILUS, Cope.

Stereochilus marginatum, Hallowell; Cope, loc. cit., 1869, 101. Georgia.

MANCULUS, Cope.

Manculus remifer. Cope, Report of Peabody Academy, Salem, Mass., 1869, p. 84. Florida.

Manculus quadridigitatus, Holbrook, N. Am. Herp., v, p. 65. North Carolina to Florida.

SPELERPES, Raf.

Spelerpes multiplicatus, Cope, Proceed. Acad. Phila., 1869, p. 106. Arkansas.

Spelerpes bilineatus, Green; Cope, loc. cit., p. 105. Eastern and Austroriparian regions, excepting Texas.

Spelerpes longicaudus, Green; Cope, loc. cit., p. 105. Eastern and Austroriparian regions, except Texas.

Spelerpes guttolineatus, Holbrook; Cope, loc. cit., p. 105. North and South Carolina, Georgia, and Alabama.

Spelerpes ruber, Daudin, subspecies *ruber*, Daudin; Cope, loc. cit., 1869, 105. Eastern and Austroriparian regions.

Spelerpes ruber, subspecies *stieticeps*, Baird, MSS. South Carolina.
Spelerpes ruber, Daudin, subspecies *montanus*, Baird; Jour. Acad. Phila., vol. i, p. 293. Alleghany Mountains, from Pennsylvania to South Carolina.

GYRINOPHILUS, Cope.

Gyrinophilus porphyriticus, Green; Cope, Proceed. Acad. Phila., 1869, p. 108. Alleghany Mountains, from New York to Alabama.

ANAIDES, Baird.

Anaides lugubris, Hallowell; Cope, loc. cit., 1869, p. 109. Entire Pacific region.
Anaides ferreus, Cope, loc. cit., 1869, p. 109. Oregon.

DESMOGNATHIDAE.

DESMOGNATHUS, Baird.

Desmognathus ochrophaea, Cope, Proceed. Acad. Phila., 1869, p. 113. Alleghany Mountains, from New York to Georgia.
Desmognathus fusca, Rafinesque; Cope, loc. cit., 115; subspecies *fusca*, Raf.; Cope, loc. cit., 116. Essex County, Massachusetts, to Biloxi, Mississippi.
Desmognathus fusca, Raf., subspecies *auriculata*, Holbrook; Cope, loc. cit., p. 116. South Carolina to Louisiana.
Desmognathus nigra, Green; Cope, loc. cit., p. 117. Alleghany Mountains, from Pennsylvania southward.

PLEURODELIDAE.

DIEMYCTYLUS, Rafinesque.

Diemyctylus torosus, Eschscholz; Girard, U. S. Expl. Exped., 1858, p. 5. Pacific region.
Diemyctylus miniatus, Raf., subspecies *miniatus*, Raf.; Hallowell, loc. cit.; Holbrook, N. Am. Herp., v, p. 57. Eastern and Austroriparian regions.
Diemyctylus miniatus, Raf., subspecies *viridescens*, Raf.; Holbrook, N. Am. Herp., v, p. 77. Eastern and Austroriparian regions.

ANURA.

BUFONIFORMIA.

BUFONIDAE.

Bufo, Laurenti.

Bufo punctatus, Baird; Girard, U. S. Mex. Bound. Surv., ii, p. 25. Sonoran and Lower Californian regions.

Bufo debilis, Girard; Baird, U. S. Mex. Bound. Surv., ii, p. 26 (*B. insidior*). Sonoran region.

Bufo halophilus, Baird; Girard, U. S. Mex. Bound. Surv., ii, p. 26. Pacific region.

Bufo columbiensis, Baird; Girard, Herpetology U. S. Expl. Exped., 77. Pacific region and Montana.

Bufo alvarius, Girard, U. S. Mex. Bound. Surv., ii, p. 26. Sonoran region.

Bufo microscaphus, Cope, Proc. Acad. Nat. Sci. Phila., 1866, p. 301. Sonoran region.

Bufo speciosus, Girard, U. S. Mex. Bound. Surv., ii, p. 26. Lower Rio Grande (Sonoran).

Bufo lentiginosus, Shaw, subspecies *frontosus*, Cope, Proc. Acad. Phila., 1866, p. 301. Sonoran region.

Bufo lentiginosus, subspecies *cognatus*, Say; Holbrook, N. Am. Herp., v, p. 21. Texan district.

Bufo lentiginosus, subspecies *americanus*, LeConte; Holbrook, Girard, U. S. Mex. Bound. Surv., ii, p. 25. Eastern and Austroriparian regions to the plains.

Bufo lentiginosus, subspecies *lentiginosus*, Latr.; Holbrook, N. Am. Herp., v, p. 7. Austroriparian region.

Bufo lentiginosus, subspecies *fowlerii*, Putnam, MSS. Massachusetts to Lake Winnipeg.

Bufo quercicus, Holbrook, N. Am. Herp., v, p. 13; Cope, Proc. Acad. Phila., 1862, p. 341. Floridan and Eastern Lousianian districts to North Carolina.

Bufo valliceps, Wiegmann; Girard, U. S. Mex. Bound. Surv., ii, p. 25, pl. xl, figs. 1–4 (*B. nebulifer*, Girard). Texan district (also Mexico).

FIRMISTERNIA.

ENGYSTOMIDAE.

ENGYSTOMA, Fitzinger.

Engystoma carolinense, Holbrook, N. Am. Herp., v, p. 23. Austroriparian region.

ARCIFERA.

HYLIDAE.

ACRIS, Dum., Bibr.

Acris gryllus, LeConte, subspecies *gryllus*, Holbrook, N. Am. Herp., iv, p. 131. Austroriparian region.
Acris gryllus, LeConte, subspecies *crepitans*, Baird, U. S. Mex. Bound. Surv., ii, p. 28. Eastern and Central regions.

CHOROPHILUS, Baird.

Chorophilus triseriatus, Wied, subspecies *clarkii*, Baird, U.S. Mex. Bound. Surv., p. 28. Texan district.
Chorophilus triseriatus, subspecies *triseriatus*, Wied. Central and Eastern regions.
Chorophilus triseriatus, subspecies *corporalis*, Cope, MSS. New Jersey.
Chorophilus nigritus, LeConte; Holbrook, N. Am. Herp., iv, p. 107. South Carolina and Georgia.
Chorophilus angulatus, Cope (*Cystignathus ocularis*), Holbrook, N. Am. Herp., iv, p. 137. South Carolina.
Chorophilus ocularis, Daudin (*Cystignathus ornatus*), Günther, Cat. Bat. Salien. Brit. Mus., p. 29. South Carolina and Georgia.
Chorophilus ornatus, Holbrook, N. Am. Herp., iv, p. 25. South Carolina; Georgia.

HYLA, Laurenti.

Hyla curta, Cope, Proc. Acad. Phila., 1866, p. 313. Lower Californian region.
Hyla regilla, Baird; Girard, U. S. Expl. Exped., p. 60. Pacific region.
Hyla eximia, Baird, U. S. Mex. Bound. Surv., p. 29. Sonoran region.
Hyla andersonii, Baird; Cope, Proc. Phila. Acad., 1862, 154. New Jersey to South Carolina.

Hyla squirella, Daudin; Holbrook, N. Am. Herp., iv, pl. 30. Austroriparian region.

Hyla carolinensis, Pennant; Holbrook, N. Am. Herp., iv, p. 29. Austroriparian region.

Hyla carolinensis, Penn., subspecies *semifasciata,* Hallowell, Proc. Acad. Phila., 1856, 306. Texan district.

Hyla pickeringii, Holbrook, N. Am. Herp., iv, pl. 34. Eastern region.

Hyla femoralis, Daudin; Holbrook, N. Am. Herp., iv, p. 31. Eastern part of Austroriparian region.

Hyla versicolor, LeConte; Holbrook, N. Am. Herp., iv, p. 28. Eastern and Austroriparian regions.

Hyla arenicolor, Cope; Baird, U. S. Bound. Surv., 29. Sonoran region.

Hyla cadaverina, Cope; Hallowell, U. S. P. R. R. Surv., x, Williamson's Report, 21. Pacific region.

Hyla gratiosa, LeConte, Proc. Acad. Phila., 1856, 146. Florida; Lower Georgia.

SMILISCA, Cope.

Smilisca baudinii, Dum., Bibr.; Baird, U. S. Bound. Surv., p. 29, pl. xxxviii, figs. 1–3. Lower Rio Grande, Mexico.

CYSTIGNATHIDAE.

LITHODYTES, Cope.

Lithodytes ricordii, Dum., Bibr.; Cope, Proc. Acad. Phila., 1862, 153. Southern Florida (Bahamas; Cuba).

EPIRHEXIS, Cope.

Epirhexis longipes, Baird, U. S. Mex. Bound. Surv., pl. xxxvii, figs 1–3. Lower Rio Grande.

SCAPHIOPIDAE.

SPEA, Cope.

Spea bombifrons, Cope, Proc. Acad. Phila., 1863, p. 53. Central region.

Spea hammondii, Baird; Cope, Proc. Acad. Phila., 1863, p. 53. Pacific region to San Diego.

Spea multiplicata, Cope, loc. cit., p. 52. Near city of Mexico.

SCAPHIOPUS, Holbrook.

Scaphiopus varius, Cope, subspecies *varius,* Cope, loc. cit., p. 52. Lower California.

Scaphiopus rarius, Cope, subspecies *rectifrenis*, Cope, loc. cit., p. 53. Sonoran region.
Scaphiopus couchii, Baird; Cope, loc. cit., p. 52. Sonoran region.
Scaphiopus holbrookii, Harlan; Cope, loc. cit., p. 54. Eastern and Austroriparian regions.

RANIFORMIA.

RANIDAE.

RANA, Linn.

Rana areolata, Baird and Girard, subspecies *capito*, LeConte, Proc. Acad. Phila., 1855, p. 425. Floridan district.
Rana areolata, Baird and Girard, subspecies *areolata*, Bd. Gir., U. S. Mex. Bound. Surv., 28, pl. xxxvi, figs. 11-12. Texan district.
Rana montezumae, Baird, U. S. Mex. Bound. Surv., p. 27. Mexican plateau.
Rana halecina, Kalm; Holbrook, N. Am. Herp., iv, p. 91; subspecies *halecina*, Hallowell, Proc. Acad. Phila., 1856, pp. 141, 250. Eastern coast-countries of Eastern and Austroriparian regions.
Rana halecina, Kalm, subspecies *berlandieri*, Baird, U. S. Mex. Bound. Surv., p. 27. Entire Interior of North America; Mexico.
Rana palustris, LeConte; Holbrook, N. Am. Herp., iv, p. 95. Eastern region.
Rana septentrionalis, Baird, Proc. Acad. Phila., 1854, p. 61 (*R. sinuata*, Bd.). Canada to Montana.
Rana clamitans, Merrem.; Holbrook, N. Am. Herp., iv, pp. 85-87. Eastern region, Louisianian district.
Rana catesbiana, Shaw; Holbrook, N. Am. Herp., iv, p. 77. Eastern and Austroriparian regions.
Rana temporaria, Linn., subspecies *aurora*, Bd.; Gird., U. S. Expl. Exped. Herp., p. 18.
Rana temporaria, Linn., subspecies *silvatica*, LeConte; Holbrook, N. Am. Herp., iv, p. 24. Eastern region.
Rana temporaria, Linn., subspecies *cantabrigensis*, Baird, Proc. Acad. Phila., 1854, p. 61. Canadian district of Eastern region to Rocky Mountains.
Rana pretiosa, Baird; Girard, U. S. Expl. Exped. Herp., p. 20. Pacific subregion.

OPHIDIA.

SOLENOGLYPHA.

CROTALIDAE.

APLOASPIS, Cope.

Aploaspis lepida, Kennicott, Proc. Acad. Phila., 1861, p. 206. Western Texas.

CROTALUS, Linn.

Crotalus pyrrhus, Cope, Proc. Phila., 1866, p. 308. Central Arizona.

Crotalus mitchellii, Cope, loc. cit., 1861, p. 293. Lower California.

Crotalus cerastes, Hallowell; Baird, U. S. Mex. Bound. Surv., vol. ii, p. 14. Arizona.

Crotalus tigris, Kennicott, U. S. Mex. Bound. Surv., vol. ii, p. 14. Arizona.

Crotalus enyo, Cope, Proc. Acad. Phila., 1861, p. 293. Lower California.

Crotalus horridus, Linn.; Holbrook, N. Am. Herp., iii, p. 9. Eastern and Austroriparian regions.

Crotalus adamanteus, Beauvois, subspecies *adamanteus*, Beauvois; Baird and Girard, N. Am. Serpents, p. 3. North Carolina to Florida.

Crotalus adamanteus, Beauvois, subspecies *atrox*, Baird and Girard, Cat., p. 5. Indian Territory and Texas to Sonora and Southern and Lower California.

Crotalus adamanteus, Beauvois, subspecies *scutulatus*, Kennicott, Proc. Acad. Phila., 1861, p. 207. Arizona.

Crotalus lucifer, Baird and Girard, Cat., p. 6. Pacific subregion; mountains of Arizona.

Crotalus polystictus, Cope, Proc. Acad. Phila., 1865, p. 191. Table land of Mexico.

Crotalus confluentus, Say; Baird and Girard, loc. cit., p. 8. Central and Sonoran regions, entering Texan district of the Austroriparian.

Crotalus molossus, Baird and Girard, Cat., p. 10. Sonoran region, entering the Texan district.

CAUDISONA, Laurenti.

Caudisona rara, Cope, Proc. Acad. Phila., 1865, p. 191. Table land of Mexico.

Caudisona miliaria, Linn.; Baird and Girard, Cat., p. 11. Austroriparian region and Sonora.

Caudisona edwardsii, Baird and Girard, Cat., p. 15. Sonoran region.

Caudisona tergemina, Say; Baird and Girard, Cat., p. 14. Eastern region west of the Alleghany Mountains; Georgia.

ANCISTRODON, Beauvois.

Ancistrodon piscivorus, Lacépède, subspecies *piscivorus*, Lacépède; Baird and Girard, Cat., 19. Austroriparian region, except Texas.

Ancistrodon piscivorus, Lacépède, subspecies *pugnax*, Baird and Girard, Cat., p. 20. Texan district.

Ancistrodon contortrix, Linn.; Baird and Girard, Cat., p. 17. Entire Eastern and Austroriparian regions.

Ancistrodon atrofuscus, Troost.; Holbrook, N. Am. Herp., iii, p. 43. Mountains of Tennessee and North Carolina.

PROTEROGLYPHA

ELAPIDAE.

ELAPS, Schneider.

Elaps fulvius, Linn., Baird and Girard, Cat., p. 21; subspecies *fulvius*. Austroriparian region.

Elaps fulvius, Linn., subspecies *tener*, Baird and Girard, Cat., p. 22. Texas.

Elaps euryxanthus, Kennicott, Proc. Acad. Phila., 1860, p. 337. Sonoran region.

Elaps distans, Kennicott, loc. cit., p. 338. Chihuahua; Florida.

ASINEA.

COLUBRIDAE.

CARPHOPHIOPS, Gervais.

Carphophiops helenae, Kennicott, Proc. Acad. Phila., 1859, p. 100. Southern Illinois; Mississippi.

Carphophiops amoenus, Say; Baird and Girard, Cat., p. 129. Massachusetts to Louisiana and Illinois.

Carphophiops vermis, Kennicott, Proc. Acad. Phila., 1859, p. 99. Missouri; Kansas.

VIRGINIA, Baird and Girard.

Virginia harperti, Dum., Bibr., Erpétologie Générale, vol. vi, p. 135. Texas; ?Georgia.

Virginia valeriae, Baird and Girard, Cat., p. 127. Maryland to Illinois and North Carolina.

Virginia elegans, Kennicott, Proc. Acad. Phila., 1859, p. 99. Southern Illinois; Arkansas.

HALDEA, Baird and Girard.

Haldea striatula, Linn.; Baird and Girard, Cat., p. 122. Virginia to Texas.

TANTILLA, Baird and Girard.

Tantilla planiceps, Blainville; Baird and Girard, Cat., p. 154. Lower California.

Tantilla gracilis, Baird and Girard, Cat., p. 132. Texas.

Tantilla hallowellii, Cope, Proc. Acad. Phila., 1861, p. 7. Texas.

Tantilla nigriceps, Kennicott, Proc. Acad. Phila., 1860, 328. Texas; New Mexico; Arizona.

Tantilla coronata, Baird and Girard, Cat., p. 131. Georgia; Mississippi.

ABASTOR, Gray.

Abastor erythrogrammus, Daudin; Baird and Girard, Cat., 125. North Carolina to Alabama.

FARANCIA, Gray.

Farancia abacura, Holbrook; Baird and Girard, Cat., p. 123. Austroriparian region.

CHILOMENISCUS, Cope.

Chilomeniscus stramineus, Cope, Proc. Acad. Phila., 1860, p. 339. Lower California.

Chilomeniscus ephippicus, Cope, Proc. Acad. Phila., 1867, p. 85. Owen Valley, California (Sonoran subregion).

Chilomeniscus cinctus, Cope, Proc. Acad. Phila., 1861, p. 303. Sonora.

CHIONACTIS, Cope.

Chionactis occipitalis, Hallowell, U. S. Pacific R. R. Survey, vol. x, Williamson's Report, p. 15. Fort Mojave, Arizona.

Chionactis occipitalis, Hallowell, subspecies *annulata*, Kennicott, U. S. Mex. Bound. Surv., vol. ii, p. 22. Colorado Desert, Arizona.

CONTIA, Baird and Girard.

Contia mitis, Baird and Girard, Cat., p. 110. Pacific region.
Contia isozona, Cope, Proc. Acad. Phila., 1866, p. 304. Utah; Arizona.
Contia episcopa, Kennicott, U. S. Mex. Bound. Surv., ii, p. 22. Texas.
Contia pygaea, Cope, Proc. Acad. Phila., 1871, p. 222. Florida.

SONORA, Baird and Girard.

Sonora semiannulata, Baird and Girard, Cat., p. 117. Sonora.

LODIA, Baird and Girard.

Lodia tenuis, Baird and Girard, Cat., p 116. Washington Territory.

GYALOPIUM, Cope.

Gyalopium canum, Cope, Proc. Acad. Phila., 1860, 243. Arizona.

CEMOPHORA, Cope.

Cemophora coccinea, Blumenbach, Baird and Girard, Cat., p. 118. Austroriparian region.

RHINOCHILUS, Baird and Girard.

Rhinochilus lecontei, Baird and Girard, Cat., p. 120. Sonoran and Southern Pacific regions.

OSCEOLA, Baird and Girard.

Osceola elapsoidea, Holbrook; Baird and Girard, Cat., p. 133. Virginia to Florida.

OPHIBOLUS, Baird and Girard.

Ophibolus doliatus, Linn., subspecies *coccineus*, Schlegel; Baird and Girard, Cat., p. 89. Florida to New Mexico; Kansas.

Ophibolus doliatus, Linn., subspecies *amaurus*, Cope, Proc. Acad. Phila., 1860, p. 258.

Ophibolus doliatus, Linn., subspecies *gentilis*, Baird and Girard, Cat., p. 90. Arkansas.

Ophibolus doliatus, Linn., subspecies *annulatus*, Kennicott, Proc. Acad. Phila., 1860, p. 329. Kansas; Arkansas and Texas.

Ophibolus doliatus, Linn., subspecies *doliatus*, Linn.; Cope, Proc. Acad., 1860, p. 256. Maryland and Virginia to Kansas; Arkansas, Louisiana, and Texas.

Ophibolus doliatus, Linn., var. *triangulus*, Boie; Baird and Girard, Cat., p. 87. From Virginia northward to Canada, Iowa, and Wisconsin.

Ophibolus multistratus, Kennicott, Proc. Acad. Phila., 1860, p. 328. Nebraska.

Ophibolus pyrrhomelas, Cope, Proc. Acad. Phila., 1866, p. 305. Arizona and California.

Ophibolus getulus, Linn., subspecies *boylii*, Baird and Girard, Cat., p. 82. Pacific and Sonoran regions.

Ophibolus getulus, Linn., subspecies *conjunctus*, Cope, Proc. Acad. Phila., 1861, 301. Lower California.

Ophibolus getulus, Linn., subspecies *splendidus*, Baird and Girard, Cat., p. 83. Sonoran region.

Ophibolus getulus, var. *sayi*, Holbrook; Baird and Girard, Cat., p. 84. United States, between the Allegheny and Rocky Mountains, from the Gulf of Mexico to Illinois.

Ophibolus getulus, Linn.; subspecies *getulus*, Linn.; Baird and Girard, Cat., p. 85. From Maryland to Florida and Louisiana, east of the Alleghenies.

Ophibolus californiae, Blainv.; Baird and Girard, Cat., p. 153. Lower California.

Ophibolus rhombomaculatus, Holbrook; Baird and Girard, Cat., p. 86. North Carolina to Georgia.

Ophibolus calligaster, Say; Cope, Proc. Acad. Phila., 1860, p. 255. Illinois to Kansas and Arkansas.

DIADOPHIS, Baird and Girard.

Diadophis punctatus, Linn., subspecies *punctatus*, Linn.; Baird and Girard, Cat., p. 112. United States and Canada, east of the plains and Texas.

Diadophis punctatus, Linn., subspecies *stictogenys*, Cope, Proc. Acad. Phila., 1860, p. 250. Texas.

Diadophis punctatus, Linn., subspecies *amabilis*, Baird and Girard, Cat., p. 113. Pacific and Sonoran regions; occasional in Texan district and Central and Eastern regions as far as Ohio.

Diadophis dysopes, Cope, Proc. Acad., 1860, p. 251. Habitat unknown.

Diadophis arnyi, Kennicott, Proc. Acad., 1859, p. 99. Illinois and Kansas.

Diadophis regalis, Baird and Girard, Cat., p. 115. Arizona; Sonora.

CONIOPHANES, Hallowell.

Coniophanes imperialis, Girard, U. S. Mex. Bound. Surv., vol. ii, p. 23. Chihuahua.

HYPSIGLENA, Cope.

Hypsiglena ochrorhyncha, Cope, Proc. Acad., 1860, 246. Lower California north to San Diego.

Hypsiglena ochrorhyncha, Cope, subspecies *chlorophaea*, Cope, loc. cit., 1860, p. 247. Arizona.

SIBON, Fitzinger.

Sibon annulatum, Linn., subspecies *septentrionale*, Kennicott, U. S. Mex. Bound. Surv., vol. ii, p. 16. Southwestern Texas.

TRIMORPHODON, Cope.

Trimorphodon lyrophanes, Cope, Proc. Acad. Phila., 1860, p. 343. Lower California and Arizona.

PHIMOTHYRA, Cope.

Phimothyra grahamiae, Baird and Girard, Cat., p. 104. Lower California and Sonoran regions to Utah and Texas.

Phimothyra grahamiae, Baird and Girard, subspecies *hexalepis*, Cope, Proc. Acad. Phila., 1866, p. 304.

Phimothyra decurtata, Cope, Proc. Acad., 1868, p. 310. Lower California.

DROMICUS, Bibron.

Dromicus flavilatus, Cope, Proc. Acad. Phila., 1871, p. 223. Coast of North Carolina.

CYCLOPHIS, Günther.

Cyclophis vernalis, DeKay; Baird and Girard, Cat., p. 108. Eastern and Austroriparian regions; rare in the latter.

Cyclophis aestivus, Linn.; Baird and Girard, Cat., p. 106. Austroriparian region, and the Eastern as far as New Jersey, Maryland, and Southern Illinois.

COLUBER, Linn. = *Scotophis*, B.&G.

Coluber emoryi, Baird and Girard, Cat., p. 157. Texas and the Mississippi Valley to Kansas and Illinois (*C. calligaster*, Kenn.; *C. rhinomegas*, Cope).

Coluber lindheimerii, Baird and Girard, Cat., p. 74. Texas and Arkansas.

Coluber vulpinus, Baird and Girard, Cat., p. 75. Massachusetts to Michigan, Kansas and northward (*C. spiloides*, D. & B.).

Coluber quadrivittatus, Holbrook; Baird and Girard, Cat., p. 80. North Carolina to Florida.

Coluber obsoletus, Say, Kennicott, Proc. Acad. Phila., 1860, p. 330; subspecies *obsoletus*, Say; Baird and Girard, Cat., p. 73. Entire Eastern United States, from Middle Texas to Massachusetts.

Coluber obsoletus, Say, subspecies *confinis*, Baird and Girard, Cat., p. 76 (*C. rubriceps*, D. & B.). Austroriparian region; Western Missouri.

Coluber guttatus, Linn.; Baird and Girard, Cat., p. 78. Austroriparian region to Central Virginia.

SPILOTES, Wagler. = *Georgia* B.&G.

Spilotes couperii, Holbrook; Baird and Girard, Cat., p. 92. Georgia.

Spilotes erebennus, Cope; Baird and Girard, Cat., p. 158. Texas to Alabama (*Georgia obsoleta*, B. & G.).

PITYOPHIS, Holbrook. — *Pituophis* B.&G.

Pityophis melanoleucus, Daudin; Baird and Girard, Cat., p. 65. New Jersey to South Carolina and Ohio.

Pityophis sayi, Schlegel, subspecies *sayi*, Schlegel; Baird and Girard, Cat., p. 151. Illinois to Kansas and northward.

Pityophis sayi, Schlegel, var. *mexicanus*, Duméril et Bibron, Erp. Gén., vol. vii, p. 236. Sonoran and Central regions, entering the Texan district.

Pityophis sayi, Schlegel, var. *bellona*, Baird and Girard, Cat., p. 66. Sonoran and Pacific regions, with Nevada and Utah.

Pityophis catenifer, Blainville; Baird and Girard, Cat., p. 69. Pacific region.

Pityophis vertebralis, Blainville; Cope, Proc. Acad. Phila., 1860, p. 342 (*P. haematois*, Cope). Lower California.

Pityophis elegans, Kennicott, U. S. Mex. Bound. Surv., p. 18. Sonora region.

BASCANIUM, Baird and Girard. = *Bascanion B+G* [handwritten]

Bascanium constrictor, Linn.; Baird and Girard, Cat., p. 93. Central, Austroriparian, and Eastern regions.

Bascanium constrictor, Linn., subspecies *vetustum*, Baird and Girard, Cat., p. 97. Pacific region.

Bascanium anthicum, Cope, Proc. Acad. Phila., 1862, p. 238. Louisiana (?).

Bascanium flagelliforme, Catesb., subspecies *flagelliforme*, Baird and Girard, Cat., p. 98. South Carolina to Florida.

Bascanium flagelliforme, Catesb., subspecies *piceum*, Cope, MS. Camp Grant, Arizona.

Bascanium flagelliforme, Catesb., subspecies *testaceum*, Say ; Baird and Girard, Cat., pp. 99 and 150. Lower Californian and Sonoran regions, with Nevada, Utah, and Texas.

Bascanium taeniatum, Hallowell, subspecies *laterale*, Hallowell, Proc. Acad. Phila., 1853. Sonoran and Pacific regions.

Bascanium taeniatum, Hallowell, subspecies *taeniatum*, Hallowell; Baird and Girard, Cat., pp. 103 and 160. Pacific and Sonoran regions; Utah and Nevada.

Bascanium taeniatum, Hallowell, subspecies *ornatum*, Baird and Girard, Cat., p. 102. Western Texas.

Bascanium aurigulum, Cope, Proc. Acad. Phila., 1861, p. 301. Lower California.

CHILOPOMA, Cope.

Chilopoma rufopunctatum, Cope, Report on Reptiles of Wheeler's Survey west of one hundredth meridian, 1875 (MS.). Sonoran district.

EUTAENIA, Baird and Girard.

Eutaenia saurita, Linn.; Baird and Girard, Cat., p. 24. Austroriparian and Eastern regions.

Eutaenia sackenii, Kennicott, Proc. Acad. Phila., 1859, p. 99. Floridan district.

Eutaenia faireyi, Baird and Girard, Cat., p. 25. Mississippi Valley, from Louisiana to Wisconsin.

Eutaenia proxima, Say ; Baird and Girard, Cat., p. 25. Valley of the Mississippi, from Wisconsin to Louisiana ; Texas ; Northeastern Mexico.

Eutaenia radix, Baird and Girard, Cat., p. 34. Central region to Lake Michigan ; Oregon.

Eutaenia macrostemma, Kennicott, subspecies *megalops*, Kennicott, Proc. Acad. Phila., 1860, p. 330. Sonoran region.

Eutaenia hammondii, Kennicott, Proc. Acad. Phila., 1860, p. 332. Pacific region.

Eutaenia mareiana, Baird and Girard, Cat., p. 36. Arkansas, Texas, and entire Rio Grande Valley.

Eutaenia vagrans, Baird and Girard, subspecies *vagrans*, Baird and Girard, Cat., p. 35. Central, Pacific, and northern parts of Sonoran regions.

Eutaenia vagrans, Baird and Girard, subspecies *angustirostris*, Kennicott, Proc. Acad. Phila., 1860, p. 332. Southern Sonoran region.

Eutaenia elegans, Baird and Girard, Cat., p. 34. California.

Eutaenia cyrtopsis, Kennicott, Proc. Acad. Phila., 1860, p. 333. Lower Californian and Sonoran regions.

Eutaenia ornata, Baird, U. S. Mex. Bound. Surv., p. 16. Valley of the Rio Grande del Norte.

Eutaenia sirtalis, Linn., subspecies *dorsalis*, Baird and Girard, Cat., p. 31. Entire North America.

Eutaenia sirtalis, Linn., subspecies *ordinata*, Linn.; Baird and Girard, Cat., p. 32. Northern part of Eastern region; Nova Scotia; North Alabama.

Eutaenia sirtalis, Linn., subspecies *sirtalis*, Linn.; Baird and Girard, Cat., p. 30. North America, excepting the Sonoran, Lower Californian, and southern half of Pacific regions.

Eutaenia sirtalis, Linn., subspecies *parietalis*, Say, Long's Exped. Rocky Mts., i, p. 186. Central and Pacific regions; Illinois.

Eutaenia sirtalis, Linn., subspecies *obscura*, Cope, MS. Eastern subregion north of Washington; northern part of Pacific region.

Eutaenia sirtalis, Linn., subspecies *dorsalis*, Baird and Girard, Cat., p. 31. North America, except the Sonoran and Lower Californian regions.

Eutaenia sirtalis, Baird and Girard, subspecies *pickeringii*, Baird and Girard, Cat., p. 29. Pacific region; Minnesota; Texas.

Eutaenia sirtalis, Linn., subspecies *tetrataenia*, Cope, MS. Pitt River, California.

Eutaenia atrata, Kennicott, Cooper and Suckley's Zoöl. Wash. Terr., p. 296. California.

Eutaenia cooperii, Kennicott, in Cooper and Suckley's Nat. Hist. Wash. Terr., p. 296. Washington and Oregon.

STORERIA, Baird and Girard.

Storeria occipitomaculata, Storer; Baird and Girard, Cat., p. 137. Eastern region; South Carolina; Georgia.

Storeria dekayi, Holbrook; Baird and Girard, Cat., p. 135. Central, Austroriparian, and Eastern regions.

TROPIDOCLONIUM, Cope.

Tropidoclonium storerioïdes, Cope, Proc. Acad. Phila., 1865, p. 190. Plateau of Mexico.

Tropidoclonium lineatum, Hallowell, Proc. Acad. Phila., 1856. Kansas to Texas.

Tropidoclonium kirtlandii, Kennicott, Proc. Acad. Phila., 1856, p. 95. Illinois; Ohio.

TROPIDONOTUS, Kuhl.

Tropidonotus clarkii, Baird and Girard, Cat., p. 48. Texas.

Tropidonotus grahamii, Baird and Girard, Cat., p. 47. The Mississippi Valley, from Louisiana to Wisconsin; Michigan.

Tropidonotus leberis, Linn.; Baird and Girard, Cat., p. 45. Austroriparian and Eastern regions, including Texas.

Tropidonotus rigidus, Say; Baird and Girard, Cat., p. 46. Pennsylvania to Georgia, east of the Alleghany Mountains.

Tropidonotus validus, Kennicott, subspecies *validus*, Kennicott, Proc. Acad. Phila., 1860, p. 334. Lower Californian and Sonoran regions; Utah.

Tropidonotus validus, Kennicott, subspecies *celaeno*, Cope, Proc. Acad. Phila., p. 341. Lower California.

Tropidonotus compsolaemus, Cope, Proc. Acad. Phila., 1860, p. 368. Florida.

Tropidonotus compressicaudus, Kennicott, Proc. Acad. Phila., 1860, p. 335. Florida.

Tropidonotus ustus, Cope, Proc. Acad. Phila, 1860, p. 340. Florida.

Tropidonotus fasciatus, Linn.; Baird and Girard, Cat., p. 39. Austroriparian region.

Tropidonotus sipedon, Linn., subspecies *sipedon*, Linn.; Baird and Girard, Cat., p. 38. Eastern and Austroriparian regions, excepting Texas.

Tropidonotus sipedon, Linn., subspecies *woodhousei*, Baird and Girard, Cat., p. 42. Texas to Missouri.

Tropidonotus sipedon, Linn., subspecies *couchii*, Kennicott, Proc. Acad., 1860, p. 335. Sonoran region.

Tropidonotus sipedon, Linn., subspecies *erythrogaster*, Shaw; Baird and Girard, Cat., p. 40. Austroriparian region, except Texas; Michigan and Kansas.

Tropidonotus taxispilotus, Holbrook; Baird and Girard, Cat., p. 43. North Carolina to Georgia.

Tropidonotus rhombifer, Hallowell; Baird and Girard, Cat., p. 43. Louisiana to Illinois and Michigan.

Tropidonotus cyclopium, Dum. et Bibron; Cope, Proc. Acad., 1861, p. 299. Florida.

HELICOPS, Wagler.

Helicops allenii, Garman, Proc. Bost. Soc. Nat. Hist., 1874, p. 92. Floridan district.

HETERODON, Beauv.

Heterodon platyrhinus, Latreille; Baird and Girard, Cat., p. 51. Entire Austroriparian and Eastern regions.

Heterodon platyrhinus, Latr., subspecies *atmodes*, Baird and Girard, Cat., p. 57. North Carolina to Georgia.

Heterodon simus, Linn., subspecies *simus*, Baird and Girard, Cat., p. 59. Austroriparian region, excepting Texas.

Heterodon simus, Linn., subspecies *nasicus*, Baird and Girard, Cat., p. 61. Sonoran and Central regions and Texas.

BOIDAE.

CHARINA, Gray.

Charina bottae, Blainv., Nouvelles Annales Mus. Hist. Nat., iii, 1834, 57. Lower Californian region.

Charina plumbea, Baird and Girard, Cat., p. 139. Pacific region; ? Nevada.

LICHANURIDAE.

LICHANURA, Cope.

Lichanura trivirgata, Cope, Proc. Acad. Phila., 1861, p. 304. Lower California.

Lichanura myriolepis, Cope, Proc. Acad. Phila., 1868, p. 2. Lower California.

Lichanura roseifusca, Cope, Proc. Acad. Phila., 1868, p. 2. Lower California.

SCOLECOPHIDIA.
STENOSTOMIDAE.
STENOSTOMA, Wagl.

Stenostoma dulce, Baird and Girard, Cat., p. 142. Sonoran region; Texas.

Stenostoma humile, Baird and Girard, Cat., p. 143. Pacific region.

LACERTILIA.
OPHEOSAURI.
AMPHISBAENIDAE.
RHINEURA, Cope.

Rhineura floridana, Baird; Cope, Proc. Acad. Phila., 1861, p. 75. Floridan district.

PLEURODONTA.
TYPHLOPHTHALMI.
ANIELLIDAE.
ANIELLA, Gray.

Aniella pulchra, Gray. Pacific region, from San Francisco southward.

LEPTOGLOSSA.
SCINCIDAE.
OLIGOSOMA, Girard.

Oligosoma laterale, Say; Holbrook, N. Am. Herp., ii, p. 133. Austroriparian region; Northwest South Carolina.

EUMECES, Wiegmann.

Eumeces septentrionalis, Baird, Proc. Acad. Phila., 1858, p. 256. Minnesota and Nebraska.

Eumeces egregius, Baird, Proc. Acad. Phila., p. 256. Florida.

Eumeces onocrepis, Cope, Report of Peabody Academy, Salem., 1869, p. 82. Florida.

Eumeces tetragrammus, Baird, Proc. Acad. Phila., 1858, 256. Lower Rio Grande.

Eumeces anthracinus, Baird, Jour. Acad. Phila., i, p. 293. Pennsylvania to Texas, in mountains.

Eumeces inornatus, Baird, Proc. Acad. Phila., 1856, p. 256. Nebraska.

Eumeces multivirgatus, Hallowell, Proc. Acad. Phila., 1857, p. 215. Central region.

Eumeces leptogrammus, Baird, Proc. Acad. Phila., 1858, p. 256. Central region.

Eumeces obsoletus, Baird and Girard, Proc. Acad. Phila., 1852, p. 129. Sonoran region, and borders of Central and Austroriparian.

Eumeces guttulatus, Hallowell; Sitgreaves's Report on Zuni, p. 113. Sonoran region and Western Texas.

Eumeces skiltonianus, Baird and Girard; Baird in Stansbury's Report Salt Lake, p. 349. Pacific region.

Eumeces fasciatus, Linn.; Holbrook, N. Am. Herp., ii, p. 117, and pp. 121, 127. Central, Austroriparian, and Eastern regions.

Eumeces longirostris, Cope, Proc. Acad. Phila., 1861, p. 313. Bermuda Islands.

LACERTIDAE.

XANTUSIA, Baird.

Xantusia vigilis, Baird, Proc. Acad. Phila., 1856, p. 255. Pacific sub-region.

TEIDAE.

CNEMIDOPHORUS, Wiegmann.

Cnemidophorus maximus, Cope, Proc. Acad. Phila., 1863, p. 104. Lower California.

Cnemidophorus grahamii, Baird and Girard, Proc. Acad. Phila., 1852, p. 128. Eastern Sonoran region.

Cnemidophorus sexlineatus, Linn.; Holbrook, N. Am. Herp., ii, p. 109. Sonoran and Austroriparian regions, to Southeast Virginia.

Cnemidophorus inornatus, Baird, Proc. Acad. Phila., 1858, p. 255. Southern Sonoran region.

Cnemidophorus octolineatus, Baird, Proc. Acad. Phila., 1858, p. 255. Southern Sonoran region.

Cnemidophorus perplexus, Baird and Girard, Proc. Acad. Phila., 1852, p. 128. Rio Grande Valley.

Cnemidophorus tessellatus, Say, subspecies *tessellatus*, Say; Baird, U. S. P. R. R. Surv., x, Beckwith's Report. p. 18. Southern Colorado.

Cnemidophorus tessellatus, Say, subspecies *tigris*, Baird and Girard; Stansbury's Report Salt Lake, p. 338. Pacific and Sonoran regions to Utah.

Cnemidophorus tessellatus, Say, subspecies *melanostethus*, Cope, Proc. Acad. Phila., 1863, p. 104. Southeast Arizona.

Cnemidophorus tessellatus, Say, subspecies *gracilis*, Baird and Girard, Proc. Acad. Nat. Sci. Phila., 1852, 128. Southeast Arizona.

VERTICARIA, Cope.

Verticaria hyperythra, Cope, Proc. Acad. Phila., 1863, p. 103. Lower California to San Diego.

DIPLOGLOSSA.

ANGUIDAE.

OPHEOSAURUS, Daudin.

Opheosaurus ventralis, Daudin; Holbrook, N. Am. Herp., ii, p. 139. Austroriparian region; Tennessee; Kansas.

GERRHONOTIDAE.

BARISSIA, Gray.

Barissia olivacea, Baird, Proc. Acad. Phila, 1858, p. 255. Southern California.

GERRHONOTUS, Wiegmann.

Gerrhonotus nobilis, Baird and Girard, Proc. Acad. Phila., 1852, p. 129. Sonora.

Gerrhonotus principis, Baird and Girard, Proc. Acad. Phila., 1852, p. 175. Northern Pacific region.

Gerrhonotus multicarinatus, Blainville (*G. formosus*), Baird and Girard, Proc. Acad. Phila., 1852, p. 175. Pacific and Lower Californian regions.

Gerrhonotus grandis, Baird and Girard, Proc. Acad. Phila., 1852, p. 176. Pacific region.

Gerrhonotus scincicaudus, Skilton, Am. Jour. Sci. Arts, 1849, p. 202. Pacific and Lower Californian regions.

Gerrhonotus infernalis, Baird and Girard; Cope, Proc. Acad. Phila., 1866, 322. Western Texas.

HELODERMIDAE.

HELODERMA, Wiegmann.

Heloderma suspectum, Cope; Baird, U. S. Bound. Surv., plate xxvi. Sonoran region.

IGUANIA.

IGUANIDAE.

HOLBROOKIA, Girard.

Holbrookia maculata, Girard, subspecies *maculata*, Girard; Stausbury's Report, 1852, p. 342. Central and Sonoran subregions.

Holbrookia maculata, Girard, subspecies *propinqua*, Baird and Girard, Proc. Acad. Phila. 1852, p. 126. Texas.

Holbrookia texana, Troschel; Baird and Girard, Proc. Acad. Phila., 1852, p. 125. Sonoran region; Western Texas.

CALLISAURUS, Blainville.

Callisaurus dracontoides, Blainv., subspecies *ventralis*, Hallowell; Sitgreave's Report Zuñi, p. 117. Sonoran region.

Callisaurus dracontoides, Blainv., subspecies *gabbii*, Cope, MS. Northern Lower California.

Callisaurus dracontoides, Blainv., subspecies *dracontoides*, Blainv., Nouv. Ann. de Mus., p. 426. Southern Lower California.

UMA, Baird.

Uma notata, Baird, Proc. Acad. Phila., 1858, p. 253. Sonora region.

SAUROMALUS, Duméril.

Sauromalus ater, Duméril; Baird, U. S. and Mex. Bound. Surv., p. 6. Sonoran region.

CROTAPHYTUS, Holbrook.

Crotaphytus colleris, Say; Holbrook, N. Am. Herp., ii, p. 79. Sonoran region; Central region to latitude 40°.

Crotaphytus wislizenii, Baird and Girard, Stansbury's Report Salt Lake, p. 340. Pacific and Sonoran regions; Nevada, Utah.

Crotaphytus reticulatus, Baird, Proc. Acad. Phila., 1858, p. 253. Western Texas.

DIPSOSAURUS, Hallowell.

Dipsosaurus dorsalis, Baird and Girard, Proc. Acad. Phila., 1852, p. 126. Lower Californian, Southern Pacific, and Sonoran regions.

UTA, Baird and Girard.

Uta thalassina, Cope, Proc. Acad. Phila., 1863, p. 104. Lower California.
Uta graciosa, Hallowell, Proc. Acad. Phila., 1854, p. 92. Pacific region.
Uta nigricauda, Cope, Proc. Acad. Phila., 1864, p. 176. Lower California.
Uta schottii, Baird, Proc. Acad. Phila., 1858, p. 253. Southern California.
Uta ornata, Baird and Girard, Proc. Acad. Phila., 1852, p. 126. Sonoran region.
Uta stansburiana, Baird and Girard, Stansbury's Report Salt Lake, p. 345. Pacific, Lower Californian, and Sonoran regions; Nevada, Utah.

SCELOPORUS, Wiegmann.

Sceloporus ornatus, Baird, U. S. Mex. Bound. Surv., p. 5. Southeastern Sonoran region.
Sceloporus jarrovii, Cope, MS., Zoöl. Wheeler's Expl. west of the 100th merid., 1875. Sonoran region (Southern Arizona).
Sceloporus poinsettii, Baird and Girard, Proc. Acad. Phila., 1852, p. 126. Sonoran region.
Sceloporus torquatus, Peale and Green, Proc. Acad. Phila., ii, p. 131. Southern Sonoran region.
Sceloporus couchii, Baird, Proc. Acad. Phila., 1858, p. 254. Southern Sonoran region.
Sceloporus marmoratus, Hallowell, Proc. Acad. Phila., 1852, p. 178. Sonoran region; Utah.
Sceloporus biseriatus, Hallowell, U. S. P. R. R. Surv., x, Williamson's Report, p. 6. ?Habitat.
Sceloporus undulatus, Harlan, subspecies *undulatus*, Harlan; Holbrook, Am. Herp., ii, p. 73. North America, except Sonoran and Lower Californian regions.

Sceloporus undulatus, Harlan, subspecies *thayerii*. Baird and Girard, Proc. Aca'. Phila., 1852, p. 127. California, Utah, New Mexico, and Rio Grande Valley.

Sceloporus consobrinus, Baird and Girard; Marcy's Report Red River, 1853, p. 237. Sonoran and Central regions; Oregon and Texas.

Sceloporus scalaris, Wiegmann, Herpetologia Mexicana, 1834, p. 52. Sonora.

Sceloporus floridanus, Baird, Proc. Acad. Phila., 1858, p. 254. Florida.

Sceloporus spinosus, Wiegmann, Herpetologia Mexicana, p. 50. Texas.

Sceloporus clarkii, Baird and Girard, subspecies *clarkii*, Baird and Girard, Proc. Acad. Phila., 1852, p. 127. Sonoran and Southern Pacific regions.

Sceloporus clarkii, Baird and Girard, subspecies *zosteromus*, Cope, Proc. Acad. Phila., 1863, p. 105. Lower California.

PHRYNOSOMA, Wiegmann.

Phrynosoma modestum, Girard, Stansbury's Rept. Salt Lake, p. 365. Sonoran region.

Phrynosoma platyrhinum, Girard, Stansbury's Rept Salt Lake, p. 361. Utah and Nevada.

Phrynosoma maccallii, Hallowell; Baird, U. S. Mex. Bound. Surv., p. 9. Desert of Gila and Colorado.

Phrynosoma regale. Girard, U. S. Mex. Bound. Surv., p. 9. Desert of Gila and Colorado.

Phrynosoma planiceps, Hallowell, Proc. Acad. Phila., 1852, p. 178. Southern Sonoran region.

Phrynosoma cornutum, Harlan; Girard, Stansbury's Rept. Salt Lake, p. 360. Texas.

Phrynosoma hernandezii, Girard, Herp. U. S. Expl. Exped., p. 395. New Mexico; Rio Grande Valley.

Phrynosoma douglassii, Bell, subspecies *ornatissimum*, Girard, Herp. U. S. Expl. Exped., 1858, p. 396. Sonoran region.

Phrynosoma douglassii, Bell, subspecies *douglassii*, Bell; Girard, Herp. U. S. Expl. Exped., p. 398. Entire Central region; Oregon and Washington.

Phrynosoma blainvillei, Gray; Girard, U. S. Expl. Exped. Herp., p. 400. Pacific region.

Phrynosoma coronatum, Blainville, Nouv. Mém. Museum, iv, p. 28. Lower California.

CYCLURA, Harlan.

Cyclura hemilopha, Cope, Proc. Acad. Phila., 1863, p. 105. Lower California.

ANOLIDAE.

ANOLIS, Merrem.

Anolis principalis, Linn.; Holbrook, N. Am. Herp., ii, 67. Austroriparian region.

Anolis cooperii, Baird, Proc. Acad. Phila., 1868. p. 254. ?California.

NYCTISAURA.

GECCONIDAE.

COLEONYX, Gray.

Coleonyx variegatus, Baird, U. S. Mex. Bound. Surv., p. 12. Sonoran region.

SPHAERODACTYLUS, Cuv.

Sphaerodactylus notatus, Baird, U. S. Mex. Bound. Surv., p. 12. Key West, Fla. (Cuba).

PHYLLODACTYLUS, Gray.

Phyllodactylus tuberculosus, Wiegmann, Nova Acta. K. L. C. Acad., xvii, p. 241. Sonoran region.

Phyllodactylus xanti, Cope, Proc. Acad. Phila., p. 102. Lower California.

DIPLODACTYLUS, Gray.

Diplodactylus unctus, Cope, Proc. Acad. Phila., 1863, p. 102. Lower California.

TESTUDINATA.

ATHECAE.

SPHARGIDIDAE.

SPHARGIS, Merrem.

Sphargis coriacea, Rondelet; Holbrook, N. Am. Herp., ii, p. 45. Atlantic coast to Massachusetts.

CRYPTODIRA.

CHELONIIDAE.

Thalassochelys, Fitz.

Thalassochelys caouana, Linn.; Holbrook, N. Am. Herp., ii, p. 33. Entire Atlantic coast.

Eretmochelys, Fitz.

Eretmochelys imbricata, Linn.; Holbrook, N. Am. Herp., ii, p. 39. Southern Atlantic coast.

Eretmochelys squamata, Linn.; Agassiz, Cont. Nat. Hist. U. S., i, p. 382. Pacific coast.

Chelonia, Brong.

Chelonia mydas, Schw.; Holbrook, N. Am. Herp., ii, p. 25. Atlantic coast south of Long Island.

Chelonia virgata, Schw.; Agassiz, Cont., i, p. 379. Pacific coast.

TRIONYCHIDAE.

Amyda, Agassiz.

Amyda mutica, Lesueur, Mém. du Mus. d'Hist. Nat., xv, p. 263. Middle and northern tributaries of the Mississippi, and the Saint Lawrence.

Aspidonectes, Wagl.

Aspidonectes ferox, Schweigger; Holbrook, N. Am. Herp., ii, p. 11 Georgia to Western Louisiana.

Aspidonectes spinifer, Lesueur, Mém. de Mus. d'Hist. Nat., xv, p. 258. Middle and northern tributaries of the Mississippi, and Saint Lawrence.

Aspidonectes asper, Agassiz, Cont. Nat. Hist. U. S., i, p. 405. Lower Mississippi tributaries.

Aspidonectes nuchalis, Agassiz, Cont. Nat. Hist. U. S., i, p. 406. Cumberland and Upper Tennessee Rivers, Tennessee.

Aspidonectes emoryi, Agassiz, Cont. Nat. Hist. U. S., i, p. 407. Texas.

CHELYDRIDAE.

Chelydra, Schw.

Chelydra serpentina, Linn.; Holbrook, N. Am. Herp., i, p. 139. From Canada to Ecuador. Wanting in the Pacific subregion.

MACROCHELYS, Gray.

Macrochelys lacertina, Schweigger; Holbrook, N. Am. Herp., i, p. 147. Tributaries of the Gulf of Mexico, from Florida to Western Texas, extending to Missouri in the Mississippi.

CINOSTERNIDAE.

AROMOCHELYS, Gray.

Aromochelys odoratus, Latreille; Holbrook, N. Am. Herp., i, p. 133. Austroriparian and Eastern subregions.

Aromochelys carinatus, Gray: Agassiz, Cont., i, p. 423. Louisianian district.

CINOSTERNUM, Wagl.

Cinosternum pennsylvanicum, Bose, subspecies *pennsylvanicum*, Bose; Holbrook, N. Am. Herp., i, p. 127. Austroriparian (? Texas) and Eastern subregions.

Cinosternum pennsylvanicum, Bose, subspecies *doubledayi*, Gray, Cat. Tort., Crocod., and Amphisb. B. M., p. 33. Southwestern United States.

Cinosternum sonoriense, LeConte, Proc. Acad. Phila., 1854, p. 183. Arizona.

Cinosternum flavescens, Agassiz, Contrib. Nat. Hist. U. S., i, p. 430. Arizona.

Cinosternum henrici, LeConte, Proc. Acad. Phila., 1854, p. 182. Sonoran subregion.

EMYDIDAE.

PSEUDEMYS. Gray.

Pseudemys rugosa, Shaw; Holbrook, N. Am. Herp., i, p. 55. New Jersey to Virginia.

Pseudemys concinna, LeConte; Holbrook, N. Am. Herp., i, pp. 119, 65. Austroriparian region (? Texas).

Pseudemys mobiliensis, Holbrook, N. Am. Herp., i, p. 71. Florida to the Rio Grande of Texas.

Pseudemys hieroglyphica, Holbrook, N. Am. Herp., i, p. 111. Middle, Western, and Gulf States.

Pseudemys scabra, Linn.; Holbrook, N. Am. Herp., i, p. 49. North Carolina to Georgia.

Pseudemys troostii, Holbrook, N. Am. Herp., i, p. 123. Valley of the Mississippi to Illinois.

Pseudemys elegans, Wied.; Holbrook, N. Am. Herp., i, p. 115. Central region and Texan district.

MALACOCLEMMYS, Gray.

Malacoclemmys geographicus, Lesueur; Holbrook, N. Am. Herp., i, p. 99. Mississippi Valley to Pennsylvania and New York.

Malacoclemmys pseudogeographicus, Holbrook, N. Am. Herp., i, p. 103. Mississippi Valley to Wisconsin and Northern Ohio.

Malacoclemmys palustris, Gmelin; Holbrook, N. Am. Herp., i, p. 87. Coast from New York to Texas.

CHRYSEMYS, Gray.

Chrysemys picta, Herm.; Holbrook, N. Am. Herp., i, p. 75. Eastern region; Louisiana, Mississippi.

Chrysemys oregonensis, Harlan; Holbrook, N. Am. Herp., i, p. 107. Central region.

Chrysemys reticulata, Bosc; Holbrook, N. Am. Herp., i, p. 59. Gulf States.

CHELOPUS, Rafinesque.

Chelopus guttatus, Schneider; Holbrook, N. Am. Herp., i, p. 81. Eastern region east of Ohio.

Chelopus muhlenbergii, Schweigger; Holbrook, N. Am. Herp., vol. i, p. 45. New Jersey and Eastern Pennsylvania.

Chelopus insculptus, LeConte; Holbrook, N. Am. Herp., i, p. 93. Eastern region east of Ohio.

Chelopus marmoratus, Baird and Girard; Hallowell, U. S. P. R. R. Surv., x, Williamson's Report, p. 3. Pacific region.

EMYS, Brong.

Emys meleagris, Shaw; Holbrook, N. Am. Herp., i, p. 39. Alleghenian district of Eastern region to Wisconsin.

CISTUDO, Flem.

Cistudo clausa, Gm., subspecies *clausa*, Gm.; Holbrook, N. Am. Herp., i, p. 31. Eastern region and Louisianian and Floridan districts.

Cistudo clausa, subspecies *triunguis*, Agass., Contrib., i, p. 445. Austroriparian region to Georgia; Eastern Pennsylvania.

Cistudo ornata, Agass., Contrib., i, p. 445. Central region.

TESTUDINIDAE.

Testudo, Linn.

Testudo carolina, Linn.; Holbrook, N. Am. Herp., i, p. 25. Austroriparian region, not north of South Carolina.
Testudo agassizii, Cooper, Proc. Calif. Acad. Sci.* Southern Pacific and Western Sonoran regions.

CROCODILIA.

CROCODILIDAE.

Alligator, Cuv.

Alligator mississippiensis, Daudin; Holbrook, N. Am. Herp., vol. ii, p. 53. Austroriparian region.

Crocodilus, Cuv.

Crocodilus americanus, Seba.; Dum. et Bib., Erp. Gén., iii, p. 119. Floridan district.

Enumeration of genera and species.

	Genera.	Species.	
BATRACHIA.			
Trachystomata	2	2	
Proteida	1	2	
Urodela	15	49	
Anura	11	48	
	— 29	—	101
REPTILIA.			
OPHIDIA.			
Solenoglypha	4	18	
Proteroglypha	1	3	
Asinea	36	109	
Scolecophidia	1	2	
	— 42	— 132	
LACERTILIA.			
Opheosauri	1	1	
Pleurodonta	22	76	
Nyctisaura	3	5	
	— 26	— 82	
TESTUDINATA.			
Athecae	1	1	
Cryptodira	16	40	
	— 17	— 41	
CROCODILIA	2	2	
		— 257	
Total species		358	

* Referred to, vol. for 1870, p. 67.

PART III.
ON GEOGRAPHICAL DISTRIBUTION
OF THE
VERTEBRATA OF THE REGNUM NEARCTICUM,
WITH ESPECIAL REFERENCE TO THE
BATRACHIA AND REPTILIA.

I.—THE FAUNAL REGIONS OF THE EARTH.

As is well known, the life of the different regions of the earth presents marked peculiarities. The differences are, in some measure, connected with the geographical and topographical relations of the continents. To each of them, peculiar divisions of animals are found to be confined; and the sum of these, or the "fauna," is found in each case to present marked characters. The districts thus marked out are the *Australian* (which includes Australia, Van Diemen's Land, New Guinea, etc.); the *Neotropical*, including South America, the West Indies, and Mexico: the *Nearctic*, or North America; the *Ethiopian*, or Africa south of the Desert of Sahara; the *Palaeotropical*, which embraces India and the adjacent islands; and, lastly, the *Palaearctic*, or Asia north of the Himalaya, Europe, and Africa north of the Great Desert. These six districts are variously related by common forms, as well as distinguished by different ones. The name of "realms" has been given to them.

The Australian realm is peculiar in the absence of nearly all types of mammalia, except the *Ornithodelphia* and the Marsupials; in the presence of various Struthious birds; in great development of the *Elapid* serpents, and absence of the higher division of both snakes and frogs (*i. e.*, *Solenoglypha* and *Raniformia*); in the existence of *Dipnoi* (*Ceratodus*) and certain Characinid fishes. On the other hand, many of the lizards and birds are of the higher types that prevail in India and Africa, viz, the *Acrodonta* and the *Oscines*.

The polar hemispheres each possess certain common forms which are not found in the other. Thus, in the southern, which is here understood as embracing the three realms called Australian, Neotropical, and

Ethiopian,* the *Sirenian* mammalia; *Struthious* birds; *Elapid* and *Peropodous* snakes; *Dipnoan*, *Chromid*, and *Characin* fishes; and *Pleurodire* tortoises, are universal, and not, or very sparsely, found in the northern. Of other groups peculiar to the Southern or Equatorial regions, the *Edentate* mammalia belong to the Neotropical and Ethiopian; the *Osteoglossid* fishes to the Neotropical, Palaeotropical, and Australian; while monkeys occur in the southern faunae, except the Australian, and in the Palaeotropical. The Ethiopian shares many peculiarities with the Northern. Thus, Insectivorous mammals, Viperine snakes, and Raniform frogs, are only found here in the southern hemisphere.

The Neotropical realm only possesses exclusively the Platyrhine monkeys and the great majority of the humming-birds. It shares with other Southern regions the Edentate and Tapiroid mammals; Struthious, Pullastrine, and Clamatorial birds; Elapid snakes; Arciferous frogs; and Characin, Chromid, Osteoglossid, and Dipnoan fishes. It has but few types of the Northern regions; these are numerous plenodont Lacertilia, the Acrodonts being entirely absent; and a few bears, deer, and Oscine birds.

The Ethiopian realm is that one which combines the prevalent features of the Palaearctic region with the southern-hemisphere types already mentioned, together with some found elsewhere only in the Palaeotropical, and a very few peculiar. The two latter classes not being mentioned elsewhere, they may be here enumerated. This region shares, with the Indian alone, the Catarrhine monkeys, the *Elephantidae, Rhinocerotidae*, and Chamaeleons. Its peculiar types are the *Lemuridae, Hippopotamidae*, and *Cameleopardalidae*, among mammals, and *Polypteridae* and *Mormyridae* among fishes.

The Northern realms of the earth agree in possessing all the earless seals; but most of its common characters are shared by India and Africa. With these regions they possess most all of the Ruminant and Insectivorous mammals, and all the Raniform frogs. The Palaearctic and Palaeotropical are very much alike, and ought probably to be united. The latter differs in possessing monkeys, elephant, rhinoceros, and tapir, *Elapid* serpents (cobras), and *Osteoglossid* fishes. In other respects, as in mammalia generally, Oscine birds and fresh-water fishes, and reptiles generally, it agrees with Northern Asia and Europe.

The Nearctic or North American realm is that with which we have here to do. It extends from the Arctic regions to a line drawn across Northern Mexico, and includes the peninsula of Lower California. It

---

* "Eogaea" of Gill, characterized in his article "On the geographical distribution of Fishes", in the "Annals and Magazine of Natural History" (4), xv, 255.

agrees in many points with the northern fauna of the Old World, and has been united with it by some authors; but its peculiar types, and those which it shares with South America, are too numerous for such an arrangement. Its relations are exhibited in the following table:

Agrees with Palaearctic in—	Differs from Palaearctic in—	
	Peculiar forms.	Neotropical forms.
Mammalia in general....		Bassarididae.
		Procyonidae.
	Antilocapra.	Megadermatidae.
	Mephitis	Dicotyles.
		Didelphys.
Birds except		Cathartidae.
		Tanagridae.
		Icteridae.
		Clamatores in general.
		Trochilidae.
	Meleagridae	Odontophorinae.
		Alligators.
		Amivid and Gerrhonotid lizards.
		Iguanid lizards.
Emyd tortoises	Chelydra	Cinosternidae.
		Solenoglyph and Elapid venomous snakes.
Raniform frogs	Scaphiopodidae.	Arcifera.
	Plethodontidae.	
	Amblystomidae.	
Diemyctylus.		
Megalobatrachus.	Trachystomata.	
	Necturus.	
	Amphiumidae.	
Percid fishes	Aphredoderidae.	Siluridae.
Cottidae.		
Haplomi.	Hypsaeidae.	
Accipenseridae.		
Spatulariidae.		
Cyprinidae	Plagopterinae.	
Gasterosteidae.	Catostomidae.	
	Amiidae.	
	Lepidosteidae.	
Petromyzon.		

The special peculiarities of the Nearctic region are then chiefly seen in the Fishes and Batrachia. In Birds and Mammals, its prominent divergences from the northern regions of the Old World are seen in the numerous representatives of forms which are characteristically South

American. Of these, the birds offer many genera peculiar to North America, while the few Mammalia are of Neotropical genera. The greatest resemblance between the North American and Palaearctic region is seen in the Mammalia. Around the Arctic regions as well as further south, several species, both of Mammalia and Birds, are identical.

Among Mollusks there is also much resemblance. *Anodonta*, *Unio*, and *Succinea* are common to both the northern faunae, but have no common species; all three greatly predominate in numbers in North America. The snails of the west coast are very European in character, but there are but few *Pupae* in the Regio Nearctica, and no *Clausiliae*, and *Bulimus* is represented by few species.

II.—NUMBER OF SPECIES.

The numbers of the Vertebrata found in the Nearctic realm are nearly as follows:

MAMMALIA:

Monotremata	0
Marsupialia	1
Edentata	1
Rodentia	139
Insectivora	28
Chiroptera	23
Cetacea	42
Sirenia	1
Hyracoidea	0
Proboscidea	0
Perissodactyla	0
Artiodactyla { Omnivora	1
{ Ruminantia	14
Carnivora { Pinnipedia	13
{ Fissipedia	46
Primates	1
	310

AVES:

Passeres { Oscines	306
{ Clamatores	33
Zygodactyli	36
Syndactyli	20

Aves—Continued.

Psittaci	1	
Accipitres	61	
Pullastrae	12	
Gallinae	22	
Brevipennes	0	
Grallae	81	
Lamellirostres	49	
Steganopodes	13	
Longipennes	71	
Pygopodes	51	
		756

Reptilia:

Crocodilia	2	
Testudinata	41	
Lacertilia	82	
Ophidia	132	
		257

Batrachia:

Anura	48	
Urodela	49	
Gymnophidia	0	
Proteida	2	
Trachystomata	2	
		101

Pisces:

Percomorphi { Pharyngognathi	12	
Labyrinthici	0	
Distegi	178	
Rhegnopteri	2	
Epilasmia	18	
Scyphobranchii	77	
Haplodoci	3	
Anacanthini	36	
Heterosomata	22	
Plectognathi	30	
Pediculati	8	
Hemibranchii	20	
Lophobranchii	7	
Synentognathi	10	
Percesoces	13	

Pisces—Continued.
Haplomi	34
Isospondyli	70
Plectospondyli	150
Scyphophori	0
Nematognathi	27
Notacanthi	0
Glanencheli	0
Ichthyocephali	0
Holostomi	0
Enchelycephali	2
Colocephali	3
Halecomorphi	2
Ginglymodi	15
Glaniostomi	30
Selachostomi	1
Holocephali	2
Plagiostomi	46
Dipnoi	0
	816
Dermopteri	8
Leptocardii	1
Total species of Vertebrata	2,249

This number is considerably below the truth, as many of the fishes, both of the ocean and of the fresh waters, remain undescribed.

It is more difficult to state the number of species of the inferior divisions of the animal kingdom. It is asserted that 8,000 species of Coleopterous insects have been discovered in the Nearctic region, and that this is probably about two-thirds of the whole. This would give 12,000 species of this the most numerous order, and the *Lepidoptera, Hymenoptera,* and *Diptera* will follow at no great distance. Probably 50,000 is below the mark as an estimate of the number of species of insects of this region. One thousand species are to be added for the remaining *Arthropoda*—say, 200 *Myriopoda*, 400 *Arachnida,* and 400 *Crustacea.* Of worms of land and water there are numerous species, the greater proportion of which are not yet known to science.

The number of the *Mollusca* and *Molluscoida* from the coasts and interior of the North American region is about 1,824, of which only 400 are marine. Of the remainder, 1,034 live in the numerous rivers and lakes, and 400 are terrestrial and air-breathers. They are distributed among the classes as follows:

CEPHALOPODA	25
PULMONATA	400
PROSOBRANCHIATA { Fresh-water	438
{ Marine	297
HETEROPODA	28
OPISTHOBRANCHIATA	53
PTEROPODA	25
SCAPHOPODA	4
LAMELLIBRANCHIATA { Fresh-water	596
{ Marine	377

MOLLUSCOIDA.

BRACHIOPODA	10
ASCIDIA	30
BRYOZOA	39

The remaining divisions of the animal kingdom may be estimated to number nearly as follows:

ECHINODERMATA (123).

	East coast.	Interior.	West coast.
HOLOTHURIDA	32	....	4
ECHINOIDEA	50	....	18
CRINOIDEA	2	....	....
ASTEROIDEA	17	....	?

COELENTERATA (144).

MEDUSAE:			
Discophora	80	....	22
Siphonophora	3	....	2
CTENOPHORA	12	....	2
POLYPI	13	....	7
HYDROIDEA	....	3	....

The divisions of *Protozoa* are well represented in our waters, but the numbers of our *Spongiida*, INFUSORIA and RHIZOPODA, have not yet been ascertained.

III.—RELATIONS TO OTHER REALMS.

It has been already remarked that several species of *Vertebrata* are common to our northern regions and Europe, Asia, etc. Thus, the wolf extends throughout the northern hemisphere; the same may be said of the fox, the ermine, and, perhaps, of the beaver. It is not improbable that our buffalo (*Bos americanus*) is a variety only of the *B. bison* of the Old World, and that the grizzly bear (*Ursus horribilis*) bears the same relation to the European brown bear (*U. arctos*). There are also certain corresponding or representative species: thus, our red fox (*Vulpes fulvus*) is nearly related to the European fox (*V. vulgaris*), and the red squirrel (*sciurus hudsonicus*) to the *S. vulgaris* of Europe. The elk and moose (*Cervus canadensis* and *Alces americanus*) respectively answer to the *C. elaphus* and *Alces europaeus*.

The majority of American deer belong to a peculiar group (*Cariacus*) mainly characteristic of the Nearctic realm; while the species of the orders *Rodentia* and *Insectivora* are mostly of characteristically distinct species or higher groups.

Among birds, similar relations prevail. The singing-birds are the most characteristic of any continent, and here we find in North America the greatest number of species, genera, and families of birds which differ from those of the Old World. Of the latter, true thrushes, swallows, shrikes, and crows occur, but in limited numbers; while the genera of finches are mostly distinct, and the vireos, tanagers, wood-warblers, Icteridae, and mock-thrushes, which form the bulk of our avifauna, do not exist in the Old World. On the other hand, starlings, flycatchers, and warblers are absent from North America.

As we direct our observation to birds of extended flight, as the *Accipitres* and water-birds, cases of identity of species of opposite continents become more frequent. This is mostly confined here, also, to the northern regions. The marsh-hawk (*Circus cyaneus*), peregrine falcon, fish-hawk, and golden eagle are examples among Falconidae. Among owls, the cases are still more numerous; such are *Nyctea nivea*, *Surnia ulula*, *Otus brachyotus*, *Strix flammea*. Some of these present geographical varieties. Corresponding species are common here, *e. g.*, the American—

> *Haliaëtus leucocephalus* to *H. albicilla* of Europe;
> ~~Falco anatum to F. vulgaris;~~
> *Falco sparverius* to *F. tinnunculus*;
> *Falco columbarius* to *F. aesalon*;
> *Bubo virginianus* to *B. maximus*;
> *Otus vilsonianus* to *O. vulgaris*;
> etc., etc.

The Nearctic realm possesses a peculiar family, the *Cathartidae* (turkey-buzzards), which the Old World lacks, but has no vultures properly so-called.

There are several wading-birds common to the two continents; and cases of identity among the ducks, gulls, and divers are relatively still more numerous. The Gallinae are, on the other hand, entirely distinct, though not without a few corresponding species.

Among lower Vertebrata, specific identity is unknown, except in one frog (*Rana temporaria*) and a few marine fishes, with one of fresh-water, the northern pike (*Esox lucius*). The numerous tortoises of North America remind one especially of Eastern Asia and India, but the western regions of our continent are as deficient in this form of animal life as the corresponding part of the Palaearctic region. *Chelydra* is peculiarly North American, and the *Cinosternidae* are Mexican in character.

The principal Crocodilian is our alligator, which presents only minor differences from the South American caimans. The lizards are all of Neotropical families, except the scines (*Eumeces*), which are found elsewhere chiefly in Africa and Australia. The genera are nearly all peculiar, or extend a short distance into the northern parts of the Neotropical, Mexico, and the West Indies. Some families have, however, a correspondence with those of the Old World, as follows: The Nearctic—

Teiidae to Lacertidae;
Gerrhonotidae to Zonuridae;
Iguanidae to Agamidae.

The *Batrachia* present relations to the Europeo-Asiatic fauna in the species of one genus (*Rana*) of frogs, and one genus (*Notophthalmus*) of salamanders. In other respects, the Nearctic batrachian fauna is highly peculiar. The cosmopolitan genus *Hyla* (tree-frogs) exists in numerous species, several of which are terrestrial. The burrowing-frogs (*Scaphiopidae*) are nearly all peculiar to this fauna. The toads are of a peculiar division of the all but cosmopolitan genus *Bufo*. The salamanders present the greatest peculiarities. The large family of *Plethodontidae* is represented by various forms, mostly terrestrial; while the genera *Desmognathus* and *Amblystoma*, each alone in its family, present curious structural modifications. To the latter belong the Siredons, or larval *Amblystomae*, which reproduce without regard to their metamorphosis, sometimes completing it and sometimes remaining unchanged.

As permanent gill-bearing *Batrachia*, *Necturus* represents the Palaearctic *Proteus*, and Siren is quite peculiar to North America. The *Amphiuma*, or snakelike Batrachia, calls to mind the similar extinct forms of the Coal-Measures; while *Protonopsis* is represented by living species in Eastern Asia, and by a fossil genus in the Miocene of Germany.

The marine fishes embrace some species which range both coasts of the North Atlantic. Such are the salmon, the haddock, the mackerel, etc., which furnish food and occupation for a numerous population on the northeastern coast. Farther south, the mullet (*Mugil albula*) is a valued food-fish, and is caught and packed in great numbers. The fishes of the Pacific coast are mostly distinct from those of the Atlantic, except a few circumpolar forms, as *Gasterosteus aculeatus*; but several (as *Gadus rachna*, Pall.) are found also on the Asiatic coast. On the warmer coasts, a few species are common to both oceans, while others exist which have a great range over several seas, noticeable among which are certain species of *Plectognathi*, particularly of *Diodon*, *Balistes*, etc.

The fresh-water fishes embrace many families characteristic of the northern hemisphere, as the cods (*Gadidae*), *Percidae* or perch, the sculpins (*Cottidae*), pike (*Esocidae*), chubs (*Cyprinidae*), the salmon, and herring, eel, sturgeon, and lamprey families. In the catfishes, the region reminds us of the tropical and southern regions; though it is a singular fact that one of our genera (*Amiurus*) is represented by single species in China.

The suckers (*Catostomidae*) are very abundant and characteristic in all fresh waters; but here, again, a single species (*Carpiodes sinensis*) has been detected in China. This is paralleled by the genus *Polyodon* (paddle-fish), of which one species is found in the Mississippi Valley, and one in the Yang tse-kiang. The most striking peculiarity of the Nearctic waters is the presence of the family of *Lepidosteidae*, or bony gars, which is represented by two genera and numerous species. No form at all resembling these exists in any other country, excepting again one species in China, and one other which is found in the adjoining Neotropical region. Not less peculiar are the species of dog-fish (*Amia*), type of the order *Halecomorphi*, which have some remote affinities with South American forms.

The relations to the Neotropical realm are in part indicated in the table on page 57. But few species are common to the Nearctic and

Southern Neotropical realms. But one mammal (the cougar, *Felis concolor*), and no reptiles, batrachians, nor fresh-water fishes, extend into Brazil; but a number of birds are permanent residents throughout both realms. These are mostly waders, as follows:

> *Rallus crepitans.*
> *Limosa fedoa.*
> *Tryngites rufescens.*
> *Actiturus bartramius.*
> *Heteroscelus breripes.*
> *Symphemia semipalmata.*
> *Ereunetes petrificatus.*
> *Aegialitis vilsonius.*
> *Nyctherodius violaceus.*

To these must be added the turkey-vulture, *Cathartes aura*. Then certain marine birds and a few fishes extend along the coasts of both regions, but their number is comparatively small.

The number of species of the Nearctic realm which occur in the Mexican region is rather greater. The red lynx and raccoon are examples of mammals, and several species of wood-warblers, vireos, and hawks represent the birds as far south as the Isthmus of Darien. The only reptiles are the snapping-tortoise and the ringed snake *Ophibolus doliatus;* the only batrachian is the *Rana halecina berlandieri*. A few other species, as *Eutaenia sirtalis*, extend for a shorter distance into the same region.

In the higher groups of the genus and family, we have greater community with the Neotropical realm. But few genera of *Batrachia* and *Reptilia* extend to its Brazilian region, but there are a few common genera of *Mammalia* (*Mephitis, Procyon, Ursus, Sciurus, Hesperomys,* and *Didelphys*), and a number of birds, especially among the lower orders, and the scansores, syndactyli, and clamatores, particularly the *Tyrannidae*. The number of genera which enter Mexico and Central America is much greater, and I select the following from the mammals, reptiles, and batrachians as these are incapable of the migrations performed by birds. Cosmopolitan genera and those common to both the American realms are omitted.

MAMMALIA.

Lynx.
Urocyon.
Putorius.
Bassaris.
Geomys.
Thomomys.
Ochetodon.
Arvicola.
Neotoma.
Sigmodon.
Cariacus.
Antilocapra.

REPTILIA.

Crotalus.
Candisona.
Ancistrodon.
Tropidoclonium.
Tropidonotus.
Eutaenia.
Trimorphodon.
Hypsiglena.
Ophibolus.
Phimothyra.
Pityophis.
Coluber.
Taatilla.
Chilomeniscus.
Cinosternum.
Chelydra.
Pseudemys.
Chelopus.
Sceloporus.
Phrynosoma.
Heloderma.
Barissia.
Gerrhonotus.
Oligosoma.
Eumeces.
Cnemidophorus.

BATRACHIA.

Amblystoma.
Spelerpes.
Spea.
Rana.

Of fishes, the common genera of the fresh waters are few. They are *Girardinus*, *Gambusia*, *Haplochilus*, and *Fundulus* of Cyprinodontidae, and *Atractosteus* of the bony gars. The southward distribution of the above genera terminates at various points; but those which belong to the Austroriparian region, as distinguished from the Sonoran, are mainly confined to the Mexican plateau. The presence of these, together with a number of peculiar forms, indicates another region of the Nearctic, which is in many respects allied to the Austroriparian. This subject will be considered in a subsequent paper.

In comparing the Nearctic realm with the West Indian region of the Neotropical, much less resemblance can be detected, especially in the Reptiles and Batrachia. The only identical species is the *Anolis principalis*, which is common to the Austroriparian region and Cuba, and there are three others of West Indian origin found in the southern part of Florida. The *Anolis* is the only reptilian genus of wide distribution in the Nearctic realm which occurs in the West Indian region. The West Indian genus *Dromicus* is represented by one species, a rare snake from the coast of North Carolina. In Batrachia, there is no community of species and none of genera, excepting in the case of the cosmopolitan genera *Bufo* and *Hyla*.

IV.—THE REGIONS.

We may now consider the variations exhibited by the component parts of the Nearctic fauna. The distribution of types indicates six principal subdivisions, which have been called the Austroriparian, Eastern, Central, Pacific, Sonoran, and Lower Californian. The Austroriparian region extends northward from the Gulf of Mexico to the isothermal of 77° F. It commences near Norfolk, Va., and occupies a belt along the coast, extending inland in North Carolina. It passes south of the Georgia Mountains, and to the northwestward up the Mississippi Valley to the southern part of Illinois. West of the Mississippi, the boundary extends south along the southern boundary of the high lands of Texas, reaching the Gulf at the mouth of the Rio Grande.

The Eastern is the most extended, reaching from the isothermal line of 77° F. north and from the Atlantic Ocean to the elevated plains west of the Mississippi River. Many of its forms extend up the bottoms of the rivers which flow to the eastward through "The Plains." The Central region extends from the limit of the Eastern as far west as the Sierra Nevada, and south on the mountains of Nevada, and along the mountains of New Mexico. The Sonoran includes parts of Nevada, New Mexico, Arizona, and Sonora in Mexico. It does not cross the Sierra Nevada, nor the Mojave desert, nor extend into the peninsula of Lower California. It sends a belt northward on the east side of the Sierra Nevada as far as, including Owen's Valley in Eastern California, latitude 37°, and enters other valleys in Nevada in the same way. It occupies the lower valley of the Rio Grande, and extends into Texas as far as the desert east of the Rio Pecos. It extends southward in Western Mexico as far as Mazatlan. The Lower Californian region occupies the peninsula of that name as far north as near San Diego.

The peculiarities of these regions are well marked. The two regions included in Eastern North America differ from all the others in the abundance of their turtles and the small number of their lizards. Prolific of life, this area is not subdivided by any marked natural barriers. Hence, though its species present great varieties in extent of range, it is not divided into districts which are very sharply defined. The warmer regions are much richer in birds, reptiles, and insects than the cooler; and as we advance northward many species disappear, while a few others are added. The natural division of the eastern part of the continent is then in a measure dependent on the isothermal lines which traverse it. In accordance with this view, the following districts have been proposed, viz: The Carolinian; the Alleghanian; the Canadian; and the Hudsonian.

The *Austroriparian region* includes the Floridan, Louisianian, and Texan districts. It possesses many peculiar genera of reptiles not found elsewhere, while the region north of it possesses none, its genera being distributed over some or all of the remaining regions. The number of peculiar species in all departments of animal life is large. It presents the greatest development of the eastern reptile life. Sixteen genera of Reptiles and eight of Batrachia do not range to the northward, while ninety-nine species are restricted in the same manner. The peculiar genera which occur over most of its area are—

LIZARDS.

Anolis.
Oligosoma.

SNAKES.

Haldea.
Cemophora.
Tantilla.
Spilotes.
Abastor.
Farancia.

TORTOISES.

Macrochelys.

CROCODILES.

Alligator.

BATRACHIA.

Engystoma.
Manculus.
Stereochilus.
Muraenopsis.
Siren.

I have omitted from this list ten genera which are restricted to one or the other of its subdivisions. The *Siren*, the *Cemophora*, the *Anolis* (chameleon), and the *Alligator*, are the most striking of the above characteristic genera. No genus of lizards is peculiar excepting *Anolis* and *Oligosoma*, which have their greatest development in other than the Nearctic continent. Among serpents, a few genera of Neotropical character extend eastward along the region of the Mexican Gulf, as far as the Atlantic coast, which are not found in any of the Northern regions; such are *Spilotes*, *Tantilla* (occurs in Lower California), and *Elaps* (also in the Sonoran). On the other hand, *Celuta*, *Virginia*, *Haldea*, and *Storeria*, embrace small serpents which it shares with the Eastern region.

This region is the headquarters of the Batrachia, especially of the tailed forms. The majority of species of the tailless genera are found here, especially of *Hyla* (tree-toads), *Rana*, and *Chorophilus*.

There are no less than nine genera of birds which do not, or only accidentally, range northward of this district. They are—

Plotus.
Tantalus.
Platalea.
Elanus.
Ictinia.
Conurus.
Chamaepelia.
Campephilus.
Helmitherus.

All these genera, excepting the last, range into South America or farther.

Among mammals, but few species and one genus (*Sigmodon*) are confined to it. *Lepus aquaticus* and *L. palustris*, the cotton-rat, the Florida *Neotoma*, etc., and a few others, are restricted by it. The fish-fauna is very similar to that of the Eastern region, under which it will be considered.

The *Eastern region* differs from the Austroriparian almost entirely in what it lacks, and agrees with it in all those peculiarities by which it is so widely separated from the Central region. No genus of mammals is found in it which does not range into the Central or other region, excepting *Condylura* (star-nosed mole); but numerous species are confined to it, not extending into the Austroriparian. These number from twenty to twenty-five. Among birds, the following genera are shared with the more southern region only: *Quiscalus, Seiurus, Oporornis, Helmitherus, Protonotaria, Parula, Mniotilta, Ortyx.* No genus of Reptiles, and but one of Batrachians (*Gyrinophilus*), is confined to this region; but it shares all it possesses with the Austroriparian. It has but three genera of lizards, viz, *Cnemidophorus, Eumeces,* and *Scelopoprus,* which are universally Nearctic. The Batrachian genera not found in the Central are—

Scaphiopus.
Gyrinophilus.
Spelerpes.
Plethodon.
Hemidactylium.
Desmatognathus.
Menopoma.
Necturus.

The characteristics of the fish-fauna of Eastern Nearctica are much more marked; two entire orders, represented by the gar (*Ginglymodi*) and dog-fish (*Halecomorphi*), are confined to it, and a series of genera of *Percidae*, embracing many species, known as *Etheostominae*, have the same range. The *Siluridae* all belong here, as well as a great majority of the genera of *Cyprinidae* and *Catostomidae*. In all of these divisions, the region is very rich in species, owing to the abundance of everflowing rivers and streams which drain it. The *Polyodontidae* (spoon-bill or paddle-fish) are not found in any of the other regions.*

The *Central* region is characterized by the general absence of forests, as compared with the Eastern. It presents two distinct divisions, each peculiar in its vegetation : the division of the plains, which extends from the eastern border to the Rocky Mountains; and the Rocky Mountain region itself, which extends to the Sierra Nevada. The former is covered with grass, and is almost totally treeless; the latter is covered with "sage-brush" (*Artemisia*), a short stout bush, which forms extensive areas of treeless brush. The grass-covered plains are the range of the bison, though it formerly sought also the tracts of grass occasionally found among the *Artemisia*. The region, as a whole, is distinguished from the Eastern by the possession of several genera of ruminating Artiodactyles, *i. e.*, *Antilocapra*, *Haplocerus*, and *Ovis*, as well as certain species of the same group, *i. e.*, *Cariacus macrotis* (Black-tailed deer) and *C. leucurus*. Other genera of mammals which distinguish it from the Eastern are *Taxidea*, *Cynomys*, *Spermophilus*, *Dipodomys*, *Perognathus*, and *Lagomys*. A few species of *Spermophilus* extend into the northwestern portion of the Eastern; while the extensive genus *Geomys* (the subterranean gophers) range over the Central subregion, and into the Western and Gulf States the Austroriparian as far as the Savannah River. A great many species of birds are peculiar to the Central region, and the following genera :

Goscoptes.
Hydrobata.
Myadestes.
Necorys.
Salpnctes.
Piciorcus.
Choviestes.
Calamspiza.
Emberagra.
Centrorcus.
Pediocetes.

* Excepting the course of the Mississppi, and perhaps the Rio Grande.

The game-birds of the Central region are larger than those of the Eastern. Such are the sage-cock, *Centrocercus urophasianus*; the *Pediocetes phasianellus*, or cock of the plains; the *Tetrao obscurus*; several ptarmigan (*Lagopus*); and *Bonasa*; the last three Palaearctic genera also.

The reptiles are not numerous, and tortoises are especially rare. Besides the genera of lizards characteristic of the Eastern district, it adds *Phrynosoma*, *Crotaphytus*, and *Holbrookia*. Among snakes, no genus is peculiar, and the moccasins and *Elaps* are wanting. There is but one, possibly two, species of rattlesnake. Batrachians are few; most of the genera of *Anura* are found, except *Hyla*. Among salamanders, the only genus is *Amblystoma*; but this is abundant, its large larvae developing in the temporary pools of many arid regions. The burrowing-frog, *Spea bombifrons*, ranges the same region, and breeds in much the same way. No genus of Batrachians or Reptiles is peculiar to the Central region.

Fishes are few in families and species, largely in consequence of the poverty of the region in rivers and streams. In the Western Colorado and the Humboldt, perch, pike, Siluridae, herring, cod, eels, gar, dogfish, and sturgeon are entirely wanting. *Cyprinidae, Catostomidae, Salmonidae*, and *Cottidae* are the only families abundant in individuals and species. The same remarks apply in great part to the Columbia River, where, however, the *Salmonidae* have a great dvelopment. These salmon are principally marine species, which ascend the river to deposit their spawn. They belong to many species, all peculiar to the region, and embrace incredible numbers of individuals.

The *Pacific* region is nearly related to te Central, and, as it consists of only the narrow district west of the Sierra Nevada, might be regarded as a subdivision of it. It, however, lacks the mammalian genera *Bos* and *Antilocapra*, and possesses certain peculiar genera of birds, as *Geococcyx* (ground-cuckoo or chaparral-cock), *Chamaea*, and *Oreortyx* (mountain-partridge). Of marin mammalia, there are several peculiar types, as the eared seals (*Otariidae*) and sea-otter (*Enhydra*). There are some genera of reptiles, *e. g Charina*, related to the Boas, *Lodia, Aniella, Gerrhonotus*, and *Xantuia*, which do not occur in the Central subregion. There are three characteristic genera of *Batrachia*, all salamanders, viz, *Anaides, Batrachops*, and *Dicamptodon*; while the Eastern genera *Plethodon* and *Diemyctlus* re-appear after skipping the entire Central district. The other tpes of Eastern *Anura* are found here, there being two species of *Hyl*

A single species of tortoise (*Chelopus marmoratus*) exists in the Pacific region.

The fresh-water fish-fauna is much like that of the Central district in being poor in types. It adds the viviparous *Pharyngognathi* of the family of *Embiotocidae*, which is represented by a number of species. The marine fauna differs from that of the east coast in the great number of species of *Salmo* and *Sebastes* and the variety of types of *Cottidae*. In its northern regions, the genus *Chirus* and allies have their peculiar habitat. The singular genus *Blepsias* (related to *Cottus*) exists on the same coast, and several valuable species of cods (*Gadus auratus*, *G. periscopus*, and *Brachygadus minutus*), with the peculiar form *Bathymaster*, belong especially to the northern coasts.

The *Sonoran* region is strongly marked among the faunae already described. It is deficient in the species of ruminating Mammalia found in the Central, and possesses a smaller number of species of mammals than any of the others. Of birds, a few genera and several species are different from those of the Central; such are *Callipepla* (partridge), ~~Glchlopsis~~, *Mitrephorus* (*Tyrannidae*), *Campylorhynchus*, and *Geococcyx*. Most of these genera occur in Mexico, and the last-named in California also. It is in Reptiles that the great peculiarity of this region appears. The following genera are not found in any of the other regions described:

LIZARDS.

Heloderma.
Sauromalus.
Uma.
Coleonyx.

SERPENTS.

Gyalopium.
Chionactis.
Sonora.
Rhinochilus.
Chilopoma.

Eight other genera of Reptilia are peculiar to this fauna and that of the Lower Californian region, under which they are enumerated. *Heloderma*, *Coleonyx*, and allies of *Gyalopium* of the above list are more largely developed in species and individuals in the Mexican region of the Neotropical realm. Every one of the five genera of serpents of the Sonoran region is characterized by a peculiar structure of the rostral plate, which is produced either anteriorly or laterally to an unusual degree; two of the genera (*Phimothyra* and *Chilomeniscus*), common to the Lower Californian region, present the same peculiarity. This region is the headquarters of the rattlesnakes, there being no less than nine species found in it, of which six are peculiar. It also possesses a majority of the species of horned toads (*Phrynosoma*); only four of the North American species being unknown there. The Testudinate fauna is very poor, possessing a few species of Nearctic character, and three *Cinosterna*, two of them of Mexican type.

The Batrachian fauna exhibits but one genus of *Urodela*, but several of the *Anura*. Appropriately to its arid character, there is but one *Rana*, but six species of toad (*Bufo*), this being the headquarters of that genus in the Regnum Nearcticum. The eastern genus *Scaphiopus* appears here, instead of the *Spea* of the other western regions. There is one species of tree-frog.

Two species of turtles of the *Cinosternidæ* have been found. The fresh-water fish-fauna is very poor, and but little known. In the Colorado River proper, the *Salmonidae* and *Cottidae* appear to be wanting, leaving only *Cyprinidae* and *Catostomidae*. A strongly-marked division of the former, the *Plagopterinae*, which embraces three genera, is mainly restricted to the Colorado River drainage, and is the most striking feature of the fish-fauna of the Sonoran region.

The *Lower Californian* region much more nearly resembles the Sonoran than the *Pacific* region. It possesses, however, many peculiar species of birds and reptiles. Scincs appear to be wanting, but other lizards abound. The following genera of reptiles have been found here, which do not occur in any other region of Nearctica:

Lizards.

Verticaria.
Diplodactylus.
Cyclura.

Snakes.

Lichanura.

These, except the last, have been found in Mexico or South America. It shares with the Sonoran only, the following:

LIZARDS.
Dipsosaurus.
Callisaurus.
Uta.
Phyllodactylus.

SNAKES.
Trimorphodon.
Hypsiglena.
Phimothyra.
Chilomeniscus.

These genera constitute the most characteristic feature of the two faunae, not occurring in any other part of North America. *Trimorphodon, Hypsiglena,* and *Phyllodactylus* are well represented in Mexico.

Of Batrachians we have, like the Sonoran, *Hyla, Scaphiopus,* and *Bufo,* but, on the other hand, *Plethodon,* as in the Pacific and Eastern. Of the fresh-water fish-fauna, nothing is known; the streams are few and small. This region extends northward to the southern boundary of California.

Among the Invertebrata, the *Mollusca* present facts of distribution similar in significance to those derived from the study of the *Vertebrata*. Thus the Eastern, the Middle, and the Pacific districts are plainly marked out in the fresh-water and land Mollusca. To the former are entirely confined the *Streptopomatidae* and the great majority of the *Unionidae,* which together constitute more than two-thirds the species of the Nearctic realm. Of land-shells, the great series of toothed snails (*Mesodontinae*), which embraces many genera and species, is almost confined to the Eastern subregion. The same is true of the snails of the group of *Gastrodontinae* and of the genera *Hyalina* and *Hygromia*. The Central subregion is characterized by its poverty in all that respects Mollusca, while several genera of land-snails are peculiar to the Pacific region, and are largely represented by species there. One hundred of the four hundred land-shells described from the Regnum Nearcticum belong to the western coast. Among snails, the genera *Aglaja, Arionta,* and *Polymita* are represented by handsome species. *Macrocyclis* and *Bin. neya* belong especially to this region.

As is to be supposed, the *Insects* indicate a greater number of subdivisions than the other animals. The fresh-water *Crustacea* have been but sparingly studied. They seem, however, to have a wide distribution; thus *Cambarus* (craw-fish) and *Artemia* are found everywhere where physical conditions are suitable.

V.—THE AUSTRORIPARIAN REGION.

V⁂. Reptiles whose distribution corresponds with the area of the Austroriparian region—24:

Trachystomata.

Siren lacertina.

Anura.

Engystoma carolinense.
Acris gryllus gryllus.
Hyla squirella.
Hyla carolinensis.

Ophidia.

Caudisona miliaria.
Ancistrodon piscivorus.
Elaps fulvius.
Haldea striatula.
Farancia abacura.
Cemophora coccinea.
Ophibolus doliatus coccineus.
Coluber obsoletus confinis.
Coluber guttatus.
Tropidonotus fasciatus.

Lacertilia.

Oligosoma laterale.
Cnemidophorus sexlineatus sexlineatus.
Opheosaurus ventralis.
Anolis principalis.

Testudinata.

Macrochelys lacertina (except Atlantic slope).
Pseudemys mobiliensis (except Atlantic slope).
Pseudemys concinna.
Testudo carolina.

Crocodilia.

Alligator mississippiensis.

As aleady remarked, this fauna is composed of the Floridan, Louisianian, and Texan districts.

The *Floridan* district contains either peculiar species of animals, or those of West Indian or South American character. The characteristic birds are chiefly of the latter character, but among reptiles the following are confined to it:

V[b]. Species confined to the Floridan district of the above—18:

Urodela.

Manculus remifer.

Anura.

Hyla gratiosa.
Lithodytes ricordii (Cuba; Bahamas).
Rana areolata capito.

Ophidia.

Elaps distans (Sonoran also).
Contia pygaea.
Eutaenia sackenii.
Tropidonotus compsolaemus.
Tropidonotus compressicaudus.
Tropidonotus ustus.
Tropidonotus cyclopium.
Helicops allenii.

Lacertilia.

Rhineura floridana.
Eumeces egregius.
Eumeces onocrepis.
Sceloporus floridanus.
Sphaerodactylus notatus (Cuba).

Crocodilia.

Crocodilus americanus (Cuba).

Of the above, the species of *Crocodilus*, *Sphaerodactylus*, and *Lithodytes* only, have been found in the Antilles. The genera of the above list which are peculiar to the Floridan district of the Nearctic fauna are—

Lithodytes.
Helicops.
Rhineura.
Sphaerodactylus.

A venomous snake, the *Elaps distans*, is common to this district and the Sonoran fauna.

Some small mammals are confined to this region also. The genera of birds that do not range north of it are—

Certhiola.

$\left.\begin{array}{l}\textit{Zenaeda}\\ \textit{Oreopelia}\\ \textit{Starnaenas}\end{array}\right\}$ Pigeons.

$\left.\begin{array}{l}\textit{Rostrhamus}\\ \textit{Polyborus}\end{array}\right\}$ Raptores.

$\left.\begin{array}{l}\textit{Aramus}\\ \textit{Audubonia}\end{array}\right\}$ Waders.

Phoenicopterus.

$\left.\begin{array}{l}\textit{Haliplana}\\ \textit{Anoüs}\end{array}\right\}$ Terns.

The *Louisianian* district possesses the peculiarities of the austroriparian fauna already pointed out, minus those of Florida and Texas. Of *Mammalia*, the genera *Alces, Mustela, Jaculus, Arctomys, Fiber*, and *Condylura* are wanting, as well as the red-squirrel, Canada lynx, gray-rabbit, etc. Its most remarkable birds are the nonpareil finch, ivory-billed woodpecker, parrakeet, etc., while its *Elaps fulvius*, or coral-snake, is one of the most beautiful of the order. A large and dangerous rattlesnake is also confined to it, viz, *Caudisona adamantea*, and the well-known moccasin *Ancistrodon piscivorus* does not range outside of its boundaries. A species of the West Indian *Dromicus* (serpents) has been found on the Atlantic coast.

V^c. Species confined to the Louisianian district—36: (E continued to the Eastern portion; W to the Western, as far as known).

Trachystomata.

Pseudobranchus striatus. E.

Proteida.

Necturus punctatus. E.

Urodela.

Amphiuma means.
Muraenopsis tridactyla. W.
Amblystoma talpoideum. E.
Amblystoma cingulatum. E.

Stereochilus marginatum. E.
Manculus quadridigitatus. E.
Spelerpes guttolineatus. E.

Anura.

Bufo lentiginosus lentiginosus.
Bufo quercicus.
Chorophilus nigritus.
Chorophilus angulatus.
Chorophilus oculatus.
Chorophilus ornatus.

Ophidia.

Crotalus adamanteus adamanteus.
Virginia harperti.
Virginia elegans. W.
Tantilla coronata.
Abastor erythrogrammus.
Osceola elapsoidea. E.
Ophibolus rhombomaculatus.
Coluber quadrivittatus. E.
Spilotes couperii. E.
Bascanium flagelliforme flagelliforme. E.
Bascanium anthicum. W.
Tropidonotus taxispilotus.
Heterodon simus simus.

Testudinata.

Aspidonectes asper. W.
Aspidonectes ferox.
Aromochelys carinatus.
Pseudemys hieroglyphica. (?)
Pseudemys scabra.
Chrysemys reticulata.
Cistudo clausa triunguis. (Penna.)

A number of the genera of the above catalogue are not yet known to extend their range into the Floridan or Texan districts, as follows:

Pseudobranchus.
Muraenopsis.
Virginia.
Abastor.
Osceola.

The genus *Virginia* occurs within the State of Texas, but whether within the Texan district is not certain, as the line separating the latter from the Louisianian district is not well known. The *Spelerpes multiplicatus*, a rare salamander from Western Arkansas, is in the same way, of uncertain reference.

The species of the following list have a peculiar range, some of them (marked E) extending beyond the borders of the Austroriparian region V^d. Species which range along the Mississippi Valley and not eastward of it—13:

Urodela.

Amblystoma microstomum (E.).

Ophidia.

Carphophiops helenae.
Virginia elegans.
Ophibolus calligaster (E.).
Coluber emoryi (E.).
Eutaenia faireyi (E.).
Eutaenia proxima.
Tropidonotus grahamii (E.).
Tropidonotus rhombifer.

Testudinata.

Macrochelys lacertina.
Pseudemys troostii.
Malacoclemmys geographica (E.).
Malacoclemmys pseudogeographica (E.).

The *Texan* district of the *Austroriparian* region is not the range of any genus not found elsewhere, but possesses the peculiar genera of the Louisianian district, many of which are represented by corresponding and peculiar species. Seventeen such species of reptiles may be enumerated, besides a salamander and a toad. Several species of mammals are also peculiar to it, *i. e.*, five rodents and two skunks. Of birds, three appear to be, so far as known, peculiar, *Ortyx texanus*, *Vireo atricapillus*, and *Milvulus forficatus*. Many Mexican birds are found on the Rio Grande, while a few enter Texas to a greater distance, as *Icterus parisorum*. The high northwestern regions of the State should be assigned to the Sonoran fauna, as the range of the two partridges (*Callipepla squamata* and *Cyrtonyx massena*) and the finch (*Peucaea cassinii*).

Several genera of mammals, birds, and reptiles exist in the Texan region, which constitute its chief claim for distinction from the Louisianian; these are—

MAMMALS.

Dicotyles (Nt.).
Bassaris (P. Nt.).

BIRDS.

Geococcyx (P. S.).

REPTILES.

Holbrookia (C. S.).
Phrynosoma (C. S. P.).
Stenostoma (Nt. P.).

None of these are peculiar: those marked (P.) being also found in the Pacific; (C.) the Central; (S.) the Sonoran; and (Nt.) the Neotropical region. Two striking species of mammals range through the Texan district, viz, the jaguar and the peccary.

V^c. Species confined to the Texan district—21:

Caducibranchiata.

Amblystoma texanum.

Anura.

Bufo valliceps (also Mexico).
Chorophilus triseriatus clarkii.
Hyla carolinensis semifasciata.
Rana areolata areolata.

Ophidia.

Crotalus adamanteus atrox.
Ancistrodon piscivorus pugnax.
Elaps fulvius tener.
Tantilla gracilis.
Tantilla hallowellii.
Tantilla nigriceps.
Contia episcopa.
Ophibolus doliatus annulatus.
Diadophis punctatus stictogenys.
Coluber lindheimerii.

Eutaenia marciana (extends W.).
Tropidonotus clarkii.
Tropidonotus sipedon woodhousei.

Lacertilia.

Holbrookia texana.
Phrynosoma cornutum.

Testudinata.

Aspidonectes emoryi.

VI.—THE EASTERN REGION.

This fauna presents four districts, viz, the Carolinian; the Alleghenian; the Canadian; and the Hudsonian. These are distinguished by the ranges of mammals and reptiles, and the breeding-places of birds. The Carolinian fauna extends in a belt north of the Louisianian, and south of the isothermal of 71°. Its northern boundary is said to extend from Long Island, south of the hill-region of New Jersey, to the southeastern corner of Pennsylvania, and thence inland. It embraces a wide belt in Maryland and Virginia, and all of central North Carolina, and then narrows very much in passing round south of the Alleghenies of Georgia. It extends north again, occupying East Tennessee, West Virginia, Kentucky, Indiana, the greater parts of Illinois and Ohio, and the southern border of Michigan. It includes also Southern Wisconsin and Minnesota, all of Iowa, and the greater part of Missouri. The Alleghanian embraces the States north of the line just described, excepting the regions pertaining to the Canadian fauna, which I now describe. This includes Northern Maine, New Hampshire, and Vermont, with the Green Mountains; the Adirondacks and summits of the Alleghany Mountains as far as Georgia. It includes Canada East and north of the lakes. The Hudsonian fauna is entirely north of the isothermal of 50°. It has great extent west of Hudson's Bay, and is narrowed southeastward to Newfoundland.

VI[a]. Species peculiar to the Eastern region—31:

Proteida.

Necturus lateralis.

Caducibranchiata.

Menopoma fuscum.
Amblystoma bicolor.

Amblystoma xiphias.
Amblystoma jeffersonianum.
Spelerpes ruber montanus.
Gyrinophilus porphyriticus.
Desmognathus ochrophaea.
Desmognathus fusca fusca.
Desmognathus nigra.

Anura.

Bufo americanus fowlerii.
Chorophilus triseriatus corporalis.
Hyla pickeringii.
Rana palustris.
Rana temporaria silvatica.
Rana temporaria cantabrigensis
Rana septentrionalis (nearly).

Ophidia.

Caudisona tergemina.
Virginia valeriae.
Ophibolus doliatus triangulum.
Cyclophis vernalis (rare south).
Coluber vulpinus.
Pityophis sayi sayi.
Storeria occipitomaculata.
Eutaenia sirtalis ordinata.
Tropidoclonium kirtlandii.

Lacertilia.

Eumeces anthracinus.

Testudinata.

Aspidonectes spinifer.
Amyda mutica.
Pseudemys rugosa.
Chelopus guttatus.
Chelopus muhlenbergii.
Chelopus insculptus.
Emys meleagris.

The *Carolinian* fauna is not so marked among reptiles as among birds. One genus of the former, *Cnemidophorus* (swift lizard), does not range north of it, with the genera *Virginia*, *Cyclophis*, *Haldea*, and *Pityophis* among serpents. Species confined in their northern range by the same limit are—

   Ophibolus doliatus doliatus.
   Ophibolus getulus.
   Tropidonotus sipedon erythrogaster.
   Pseudemys rugosa.
   Malacoclemmys palustris.
   Hyla andersonii.

Genera of birds restricted in the same way are—

   Guiraca.
   Helmitherus.
   Mimus.
   Polioptila.
   Gallinula.
   Herodias.
   Florida.
   Himantopus.
   Recurvirostra.

The *Alleghanian* district includes nearly all of the remaining species of Reptiles and several Batrachians. The genera of these which do not extend north of it are the following:

   LIZARDS.

   Sceloporus.
   Eumeces.

   SNAKES.

   Carphophiops.
   Coluber.
   Cyclophis.
   Tropidonotus.
   Ophibolus.
   Heterodon.
   Caudisona.
   Crotalus.
   Ancistrodon.

BATRACHIA.
Chorophilus.
Hyla.
Hemidactylium.
Desmognathus.
Menopoma.
Necturus.

The species thus restricted number twenty-six. The genera of birds which do not range north of this fauna are numerous. They are—

Sialia.
Vireo.
Pyranga.
Harporhynchus.
Troglodytes.
Cyanospiza.
Pipilo.
Ammodromus.
Sturnella.
Icterus.
Zenaedura.
Cupidonia.
Ortyx.
Meleagris.
Ardetta.
Rallus.

The catamount, red-squirrel, jumping-mouse, gray-rabbit, star-nosed mole, and elk, do not range south of this fauna.

The *Canadian* fauna is distinguished for its few reptiles (there being only seven species) and Batrachia, as follows:

TORTOISES.

Chelydra serpentina.
Chelopus insculptus.
Chrysemys picta.

SNAKES.

Bascanium constrictor.
Eutaenia sirtalis.
Diadophis punctatus.
Storeria occipitomaculata.

FROGS.

Rana temporaria cantabrigensis.
Rana septentrionalis.

SALAMANDERS.

Desmognathus ochrophaea.
Desmognathus nigra.
Spelerpes ruber.
Spelerpes bilineatus.
Spelerpes longicauda.

This fauna extends south along the crests of the Alleghenies, where we find the catamount, snow-bird, red-squirrel, and brook-trout (*Salmo fontinalis*), and *Desmognathus ochrophaea*, as far as Georgia.

Several mammals are restricted in northward range by the boundary of this fauna; such are the buffalo, raccoon, skunk, wild-cat, panther, star-nosed mole, etc.; and the moose, caribou, wolverine, and fisher do not range, according to J. A. Allen, south of it.

VI^b. Species confined to the Canadian district, or nearly so:

Urodela.
Amblystoma jeffersonianum laterale.

Anura.
Bufo lentiginosus fowlerii.
Rana septentrionalis.
Rana temporaria cantabrigensis.

In the *Hudsonian* district there are no reptiles, and the fresh waters begin to present various new species of *Salmo* and *Coregonus* (trout and white-fish). The catamount, fisher, ermine, black-bear, red-squirrel, ground-hog, etc., do not range north of it. The following singing-birds breed there:

Anthus ludovicianus.
Saxicola oenanthe.
Ampelis garrula.
Aegiothus linaria.
Plectrophanes lapponica.
Plectrophanes nivalis.
Plectrophanes picta.
Leucosticte tephrocotis.

The first and last two are the only species not also found in Europe. Numerous waders and swimming-birds breed in this region, the whole number being thirty-six; while ninety-six species of birds do not wander north of it. To this category many of the common species of the Middle States belong.

North of this the species of vertebrates are circumpolar or arctic. The ichthyological fauna of the two Eastern subregions remains to be considered. For the present, they will be united, though the distribution of fresh-water fishes is governed by laws similar to those controlling terrestrial vertebrates and other animals, in spite of the seemingly confined nature of their habitat. With this general principle in view, we may revert briefly to this distribution over this district of the Nearctic region. This large area is characterized by the distribution of several species in all its waters, or nearly so, so far as yet examined—those of *Semotilus, Ceratichthys, Hypsilepis, Catostomus*, etc., or by the universal recurrence of the same in suitable situations; and by the representation of these and other genera by nearly allied species in its different portions. The fauna of the tributaries of the Mississippi constitutes, it might be said, that of our district; while the slight variations presented by the Atlantic coast streams might be regarded as exceptional. The fauna of the great lakes combines the peculiarities of both, possessing as a special peculiarity, (I), which belongs to the Lake region, which, in the district, commences at latitude 42° and extends to the Arctic regions, the range of the genus Coregonus. The peculiarity of the Atlantic subdistrict (II) may be said to be the abundance of *Esox, Salmo*, and *Anguilla*, and the absence of *Haploidonotus*. The first two are abundant in the Lake region, while *Anguilla* and *Haploidonotus* have but a partial distribution there. In (III), the Mississippi basin, *Esox* is represented by but few species, and remarkably few individuals. *Salmo* occurs abundantly in the upper parts of the Missouri tributaries, exists in the western mountain-streams of the Alleghanies, becoming rare in those of the Kanawha, and only occurring near the highest summits in those of the Tennessee, south to the line of South Carolina. It is especially characterized by the paddle-fish (*Spatularia* or *Polyodon*), the shovel-sturgeon (*Scaphirhynchops*), and the alligator-gar (*Atractosteus*); also by the buffalo (*Bubalichthys*), the *Cycleptus*, etc., among suckers, and the fork-tailed catfish (*Ichthaelurus*). Among Percomorphs, the *Haploidonotus* is the characteristic genus; and among those allied to the herring, the genus *Hyodon*. Numerous species are confined to its affluents. The gradation from the Mississippi grouping of species to the Atlantic is very gradual, and takes place in successional order from those emptying into the Gulf of Mexico toward the east and northeast, until we reach the rivers of Massachusetts and Maine, where the greatest modification of the fauna exists. The latter fact has been pointed out by Agassiz, who calls this region a "zoölogical island," and enumerates the characteristic Nearctic genera which are wanting there. I give now a list showing the points at which Mississippi genera cease, as we follow the rivers of the Gulf and Atlantic coasts, so far as our present knowledge extends.

Gulf rivers : *Haploidonotus* has not yet been indicated from eastward of these, except in the Lake area.

Roanoke : *Campostoma* ceases here.

James : *Micropterus* and *Ambloplites* cease.

Potomac : *Pomoxys*, according to Professor Baird (verb. comm.), ceases here.

Susquehanna : *Ceratichthys, Exoglossum, Chrosomus, Carpiodes*, cease.

Delaware : *Clinostomus, Hypsilepis analostanus, Enneacanthus*, and *Lepidosteus* cease.

Hudson : *Semotilus corporalis*, according to F. W. Putnam (verb. comm.), ceases.

The types remaining in the Atlantic waters of the New England district (IV) are first, then, *Salmo, Esox, Anguilla, Perca* ; and, secondly, the general types *Boleosoma. Semotilus, Hypsilepis, Stilbe, Hybopsis* (*bifrenatus*), *Fundulus*, and *Amiurus ;* and the Lake types *Lota* and *Coregonus*.

VII.—THE CENTRAL REGION.

VII^a. Species peculiar to the Central region—12 :

Anura.

Spea bombifrons.

Ophidia.

Ophibolus multistratus.
Eutaenia radix.
Eutaenia vagrans vagrans.
Eutaenia sirtalis parietalis.

Lacertilia.

Eumeces septentrionalis.
Eumeces inornatus.
Eumeces multivirgatus.
Holbrookia maculata maculata.
Phrynosoma douglassii douglassii.

Testudinata.

Pseudemys elegans.
Chrysemys oregonensis.
Cistudo ornata.

VIII.—THE PACIFIC REGION.

VIII[a]. Species confined to the Pacific region—44:

Urodela.

Amblystoma macrodactylum.
Amblystoma paroticum.
Amblystoma tenebrosum.
Amblystoma aterrimum.
Dicamptodon ensatus.
Batrachoseps attenuatus.
Batrachoseps nigriventris.
Batrachoseps pacificus.
Plethodon intermedius.
Plethodon oregonensis.
Anaides lugubris.
Anaides ferreus.
Diemyctylus torosus.

Anura.

Bufo halophilus.
Hyla regilla.
Hyla cadaverina.
Spea hammondii.
Rana temporaria aurora.
Rana pretiosa.

Ophidia.

Crotalus lucifer.
Contia mitis.
Lodia tenuis.
Pityophis catenifer.
Bascanium constrictor vetustum.
Eutaenia hammondii.
Eutaenia elegans.
Eutaenia sirtalis pickeringii.

Eutaenia sirtalis concinna.
Eutaenia sirtalis tetrataenia.
Eutaenia cooperii.
Eutaenia atrata.
Charina plumbea.
Stenostoma humile.

Lacertilia.

Aniella pulchra.
Eumeces skiltonianus.
Xantusia vigilis.
Barissia olivacea.
Gerrhonotus principis.
Gerrhonotus grandis.
Gerrhonotus scincicaudus.
Uta graciosa.
Uta schottii.
Phrynosoma blainvillei.

Testudinata.

Chelopus marmoratus.

Gerrhonotus multicarinatus is common to the Pacific and Lower California regions.

IX.—THE SONORAN REGION.

IX^a. Species confined to the Sonoran region—68:

Anura.

Bufo alvarius.
Bufo debilis.
Bufo microscaphus.
Bufo speciosus.
Bufo lentiginosus frontosus.
Hyla eximia. (Mexico also.)
Hyla arenicolor.
Scaphiopus varius rectifrenis.
Scaphiopus couchii.

Ophidia.

Crotalus pyrrhus.
Crotalus cerastes.
Crotalus tigris.

Crotalus adamanteus scutulatus.
Crotalus molossus.
Caudisona edwardsii.
Elaps euryxanthus.
Chilomeniscus ephippicus.
Chilomeniscus cinctus.
Chionactis occipitalis.
Contia isozona.
Sonora semiannulata.
Gyalopium canum.
Rhinochilus lecontei.
Ophibolus pyrrhomelus.
Ophibolus getulus splendidus.
Diadophis regalis.
Hypsiglena ochrorhyncha chlorophaea.
Phimothyra grahamiae.
Bascanium flagelliforme piceum.
Chilopoma rufipunctatum.
Eutaenia macrostemma.
Eutaenia vagrans angustirostris.
Tropidonotus validus validus.
Tropidonotus sipedon couchii.
Stenostoma dulce.

Lacertilia.

Eumeces obsoletus.
Eumeces guttulatus.
Cnemidophorus grahamii.
Cnemidophorus inornatus.
Cnemidophorus octolineatus.
Cnemidophorus tessellatus gracilis.
Cnemidophorus tessellatus melanostethus.
Gerrhonotus nobilis.
Gerrhonotus infernalis.
Heloderma suspectum.
Callisaurus dracontoides ventralis.
Uma notata.
Sauromalus ater.
Crotaphytus reticulatus.
Uta ornata.

Sceloporus ornatus.
Sceloporus jarrovii.
Sceloporus poinsettii.
Sceloporus torquatus.
Sceloporus couchii.
Sceloporus marmoratus.
Sceloporus clarkii.
Phrynosoma modestum.
Phrynosoma maccallii.
Phrynosoma regale.
Phrynosoma planiceps.
Phrynosoma hernandezii.
Coleonyx variegatus.
Phyllodactylus tuberculatus.

Testudinata.

Cinosternum sonoriense.
Cinosternum henrici.
Cinosternum flavescens.
Testudo agassizii.

Phrynosoma platyrhinium has as yet been observed in Nevada only.

X.—THE LOWER CALIFORNIAN REGION.

X[a]. Species peculiar to the Lower Californian region—27:

Urodela.

Plethodon croceater.

Anura.

Hyla curta.

Ophidia.

Crotalus enyo.
Crotalus mitchellii.
Tantilla planiceps.
Chilomeniscus stramineus.
Ophibolus californiae.
Ophibolus getulus conjunctus.
Hypsiglena ochrorhyncha ochrorhyncha.
Phimothyra decurtata.
Pityophis vertebralis.
Bascanium aurigulum.

Tropidonotus validus celaeno.
Charina bottae.
Lichanura trivirgata.
Lichanura myriolepis.
Lichanura roseofusca.

Lacertilia.

Phyllodactylus unctus.
Phyllodactylus xanti.
Cnemidophorus maximus.
Verticaria hyperythra.
Callisaurus draconotoides.
Uta thalassina.
Uta nigricauda.
Sceloporus clarkii zosteromus.
Phrynosoma coronatum.
Cyclura hemilopha.

XI.—RELATION OF DISTRIBUTION TO PHYSICAL CAUSES.

The first observation with regard to the Batrachian and Reptilian fauna of North America is the usual one, viz., that the number of specific and generic types exhibits a rapid increase as we approach the tropics. Of the area inhabited by these forms of animals, less than one-fourth is included in the three Southern regions—the Austroriparian, the Sonoran, and the Lower Californian; yet these contain more than half of the entire number of species, and all but eight of the genera are found in them. Of this number, forty-two genera, or one-third of the total, is confined to within their boundaries. It is a truism directly resulting from the very small production of animal heat by these animals, that temperature, and therefore latitude, has the greatest influence on their life and distribution. This is exhibited in other ways than in multiplication of forms. It is well known, that although plainly-colored reptiles are not wanting in the tropics, brilliantly-colored species are much more abundant there than in temperate regions. Although the Regnum Nearcticum does not extend into the tropics, its southern districts are the habitat of most of the species characterized by bright colors. This is most instructively seen in species having a wide range. Such is the case with the southern subspecies of *Desmatognathus* among salamanders, and *Hyla* among frogs. So with snakes of the genera *Crotalus, Caudisona, Ophibolus, Bascanium,* and *Eutaenia.* It is also true of the lizards of the genera *Phrynosoma*, *Holbrookia*, and *Sceloporus*. *Eutaenia* and *Sceloporus* become metallic in the Mexican subregion, as is also the case with the Anoles. The North American species of *Anolis* does not display metallic luster, while a large part of those of Mexico and a smaller proportion of those of the West Indies exhibit it.

Another important influence in the modification of the life in question is the amount of terrestrial and atmospheric moisture. In the case of the Batrachia, this agent is as important as that of heat, since a greater or less part of their life is, in most species, necessarily spent in the water. The reptiles are less dependent on it, but, as their food consists largely of insects, and as these in turn depend on vegetation for sustenance, the modifying influence of moisture on their habits must be very great.

The Central region combines the disadvantages of low temperature, due to its elevation above the level of the sea, and of arid atmosphere; hence its poverty in *Batrachia* and *Reptilia*. There are but nine species of both classes peculiar to it, while a few others enter from surrounding areas.

The distribution in the other regions is evidently dependent on the same conditions. Thus the well-watered, forest-covered Eastern and Austroriparian regions are the home of the salamanders, the frogs, the tree-toads, and the turtles. The dry and often barren Sonoran and Central regions abound in the lizards and the toads. The Pacific region, which is intermediate in climatic character, exhibits a combination of the two types of life; it unites an abundant lizard-fauna with numerous frogs and salamanders, while there is but one tortoise.

Another character of the reptilian life of arid regions is to be seen in a peculiarity of coloration. This, which has been already observed by the ornithologists, consists of a pallor, or arenaceous hue of the body, nearly corresponding with the tints of dry or sandy earth. This prevails throughout the Batrachia and Reptilia of the Sonoran region, although it is often relieved by markings of brilliant color, of which red is much the most usual. This peculiarity doubtless results immediately from the power of metachrosis, or color-change, possessed by all cold-blooded Vertebrata, by means of which they readily assume the color of the body on which they rest. That a prevalent color of such bodies should lead to a habit of preference for that color is necessary, and as such habits become automatic, the permanence of the color is naturally established.

Another peculiarity of the Sonoran region, and which it shares with a part of Mexico, is the predominance of snakes which possess an extraordinary development of the rostral shield either forward or outward. This has also been observed by Professor Jan, who referred such genera to a group he termed the *Probletorhinidae*, but which has not sufficient definition to be retained in the system. Of ten genera of snakes in the Nearctic region which possess the character, nine are found in the Sonoran subregion, five are peculiar to it, and it shares two with the Lower Californian subregion only. One of the latter (*Phimothyra*) is closely imitated by a genus (*Lytorhynchus*) which occurs on the borders of the African Sahara. The *Heterodon* of the Eastern States, though not confined to the sandy coast-regions, greatly abounds there; and the South American species skip the forest-covered Amazon Valley and reappear on the plains of the Paraguay and Parana. As the Sonoran region embraces a number of desert areas, it is altogether probable that the peculiar forms in question have a direct relation to the removing of dry earth and sand, in the search for concealment and food. A modification of foot-structure, supposed to have relation to the same end, is seen in the movable spines on the outer side of the foot in the genus *Uma*, a character exhibited in higher perfection in the South African genus *Ptenopus*.*

The abundance of Bufones is doubtless due in part to their adaptation to life in dry regions. They are mostly furnished with tarsal bones especially developed for excavating purposes.

---
* Proc. Acad. Phila., 1868, p. 321.

PART IV.
BIBLIOGRAPHY.

The present list only includes the titles of works and memoirs which embrace discussions of systematic or distributional relations of the reptiles of the Regio Nearctica. Those embracing descriptions of species only will be added at a future time.

The subject of general geographical distribution has been especially investigated by Sclater, Huxley, and the writer; while Baird, Agassiz, LeConte, Verrill, Allen, and the writer have devoted themselves especially to the distribution of the animals of the fauna Nearctica. In 1856, Dr. Hallowell remarked the rarity of salamanders and turtles in the Sonoran region,* and Professor Baird has especially demonstrated the complementary relation exhibited in the distribution of lizards and turtles in North America. Professor Verrill and J. A. Allen have defined the faunal subdivisions of Eastern North America with great success, basing their conclusions on the distribution of birds and *Mammalia*. The writer subsequently defined the Sonoran and Lower Californian regions, and elevated the Austroriparian area to the same value, adopting, also, the districts of Verrill and Allen. In the present essay I am greatly indebted to the learned work of J. A. Allen for information on the distribution of birds, as well as to the previous essay of Professor Baird on the birds and mammals.

A.— *Works on the classification of Batrachia and Reptilia.*

1817. Cuvier. Règne Animal. First edition. Paris.
1820. Merrem. Systema Amphibiorum.
1824. Wagler, in Spix Serpentes Brazilium.
1825. Latreille. Familles Naturelles du Règne Animal. Paris.
1825. Gray. Genera of Reptiles in Annals of Philosophy. London.
1826 (June). Fitzinger. Neue Classification der Reptilien.

---
* Proc. Acad. Phila., 1856, p. 309.

1826 (October). Boie, H. Erpetologie von Java in Ferrusac's Bulletin des Sciences Naturelles et Géologiques.
1827. Boie, F., in Isis von Oken, p. 508.
1830. Wagler. Natürliches System der Amphibien.
1831. Müller. Beiträge zur Anatomie der Amphibien. Tiedemann u Treviranus' Zeitschrift für Physiologie, iv, p. 199.
1832. Wiegmann und Ruthe. Handbuch der Zoologie. Berlin.
1832. Bonaparte. Saggio di una Distribuzione Metodica degli Animali Vertebrati. Rome.
1834. Duméril et Bibron. Erpétologie Générale, vol. i. General Classification and Anatomy. The *Testudinata*. Paris.
1834. Wiegmann. Herpetologia Mexicana. Berlin.
1835. Duméril et Bibron. Erpétologie Générale, vol. ii. *Testudinata* : *Lacertilia*, in general.
1836. Duméril et Bibron. Erpétologie Générale, vol. iii. *Crocodilia*, *Chamaeleontidae*, *Geeconidae*, *Varanidae*.
1837. Duméril et Bibron. Erpétologie Générale, vol. iv. Sauriens (*Iguanidae* and *Agamidae*). Paris.
1837. Schlegel. Essai sur le Physionomie des Serpens. Hague.
1839. Duméril et Bibron. Erpétologie Générale, vol. v. *Lacertidae*, *Chalcididae*, and *Scincidae*.
1841. Duméril et Bibron. Erpétologie Générale, vol. viii. *Batrachia Gymnophiona*, and *Anura*.
1843. Fitzinger. Systema Reptilium. Vienna.
1844. Duméril et Bibron. Erpétologie Générale, vol. vi. *Ophidia* in general; *Scolecophidia* and *Asinea*, parts.
1844. Gray. Catalogue of Tortoises, Crocodiles, and Amphisbaenians in the British Museum. London.
1845. Gray. Catalogue of the Lizards in the British Museum. London.
1849. Gray. Catalogue of Specimens of Snakes in the British Museum. London.
1849. Baird. Revision of the North American Tailed Batrachia, etc. Journal of Academy, Philadelphia, vol. i, p. 281.
1850. Gray. Catalogue of the Specimens of Amphibia in the British Museum. London.
1853. (January). Baird and Girard. Catalogue of the Serpents of North America. Washington.
1853. Duméril. Prodrome de la Classification des Reptiles Ophidiens Institut de France.

1854. Duméril et Bibron. Erpétologie Générale. Tome vii, part 1, *Ophidia Asinea;* part 2, Venomous Serpents. Tome ix, *Batrachia Urodela.* Tome x, plates.
1854. LeConte, J. Catalogue of the North American *Testudinata.* Proceedings of Philadelphia Academy, vol. vii.
1855. Gray. Catalogue of the Shield Reptiles in the British Museum. London.
1857. Agassiz. Contributions to the Natural History of the United States, part ii. North American *Testudinata,* p. 233.
1858. Gray. On the Classification of the Old World Salamanders. Proceedings of the Zoölogical Society, London, p. 235.
1858. Günther. Catalogue of the Colubrine Serpents in the British Museum. London.
1858. Günther. Catalogue of the Batrachia Salientia in the British Museum. London.
1859. Cope. Catalogue of the Venomous Serpents. Proceedings of the Academy, Philadelphia, 1859, p. 330.
1860. Owen. Paleontology. London. (Arrangement of Extinct Reptiles.)
1863. Jan. Elenco Sistematico degli Ofidi Descritti e Disegnati per l'Iconografia Generale. Milan.
1864. Cope. Characters of the Higher Groups of Reptilia Squamata. Proceedings of the Academy, Philadelphia, p. 224.
1864. Günther. Reptiles of British India. Ray Society.
1865. Cope. Sketch of the Primary Groups of Batrachia Salientia Natural History Review. London.
1866. Cope. On the Arciferous *Anura* and the *Urodela.* Journal of the Academy of Natural Sciences, Philadelphia.
1867. Cope. On the Families of the Raniform *Anura.* Journal of the Academy, Philadelphia, p. 189.
1867. Günther. Contribution to the Anatomy of *Hatteria.* Philosophical Transactions.
1869. Cope. Synopsis of the Extinct Batrachia, Reptilia, and Aves of North America. Transactions of the American Philosophical Society, vol. xiv.
1869. Cope. A Review of the Species of *Plethodontidae* and *Desmognathidae.* Proceedings of the Academy, Philadelphia, p. 93.
1870. Cope. On the Homologies of some of the Cranial Bones of the Reptilia, and on the Systematic Arrangement of the Class. Proceedings of the American Association for the Advancement of Science, p. 194. Cambridge.

1870. Gray. Supplement to the Catalogue of Shield Reptiles in the British Museum. London.
1872. Huxley. Anatomy of the Vertebrata. London.

B.—*Works treating of the geographical distribution of North American Batrachia and Reptilia.*

1857. Agassiz. Contributions to the Natural History of the United States, vol. i, part i, p. 449. On the Geographical Distribution of North American Testudinata.
1866. Baird. The Distribution and Migration of North American Birds. American Journal of the Sciences and Arts, p. 78, 184–347 (January).
1866. Verrill. Report of some Investigations upon the Geographical Distribution of North American Birds. Proceedings of the Boston Society of Natural History, vol. x, p. 259 (May).
1866. Cope. On the *Reptilia* and *Batrachia* of the Sonoran Province of the Nearctic Region. Proceedings of the Philadelphia Academy, p. 300 (October).
1869. Cope. On the Origin of Genera. Philadelphia.
1871. Allen, J. A. Bulletin of the Museum of Comparative Zoölogy. vol. ii, No. 3, p. 404.
1873. Cope. Gray's Atlas of the United States, p. 32. Geographical Distribution of North American Vertebrata (with map).

INDEX.

	Page.		Page.
Abastor	35	Bufo	29
Achrochordidae	22	Bufonidae	9, 29
Acontiidae	20	Bufoniformia	9, 29
Acris	30	Caducibranchiata	25
Adocidae	17	Caeciliidae	11
Agamidae	18	Callisaurus	47
Aglossa	9	Canadian fauna	85
Alleghanian district	84	Carolinian fauna	84
Alligator	54	Carphophiops	34
Amblystoma	25	Candisona	33
Amblystomidae	12, 25	Causidae	23
Amphicoelia	14	Cemophora	36
Amphisbaenidae	20, 44	Central region	71, 88
Amphiuma	25	Chalcidae	19
Amphiumidae	12, 25	Chamaeleontidae	17
Amyda	51	Charina	43
Anaides	28	Check-list of the species of Batrachia	
Ancistrodon	34	and Reptilia of the Nearctic or	
Anguidae	18, 46	North American realm	24
Aniella	44	Cheloniidae	16, 51
Aniellidae	20, 24	Chelonia	51
Anolidae	18, 50	Chelopus	53
Anolis	50	Chelydidae	17
Anomodontia	15	Chelydra	51
Anthracosauridae	10	Chelydridae	16, 51
Anura	7, 29	Chilomeniscus	35
Aploaspis	33	Chionactis	35
Areifera	9, 30	Chilopoma	40
Aromochelys	52	Chorophilus	30
Arrangement of the families and higher divisions of Batrachia and Reptilia	7	Chrysemys	53
		Cinosternidae	16, 52
		Cinosternum	52
Aspidonectes	51	Cistudo	53
Asinea	21, 34	Cnemidophorus	45
Asterophrydidae	10	Cocytinidae	12
Athecae	16, 50	Coleonyx	50
Atractaspididae	23	Colosteidae	10
Austroriparian region	68, 76	Colostethidae	7
Baphetidae	10	Coluber	39
Barissia	46	Colubridae	22, 34
Bascanium	40	Compsognathidae	13
Batrachia	7, 24	Coniophanes	38
Batrachophrynidae	9	Contia	36
Batrachoseps	26	Crocodilia	14, 54
Belodontidae	14	Crocodilidae	14
Bibliography	97	Crocodilus	54
Boidae	22, 43	Crotalidae	23, 33
Brevicipitidae	8	Crotalus	33

	Page.		Page.
Crotaphytus	47	Helodermidae	19, 47
Cryptodira	16, 51	Hemidactylium	26
Cyclophis	38	Hemiphractidae	10
Cyclura	50	Hemisidae	-
Cystignathidae	9, 31	Heterodon	43
Dactylethridae	9	Holbrookia	47
Dendrobatidae	2	Homalopsidae	22
Desmognathidae	11, 28	Hudsonian district	56
Diadophis	37	Hydraspididae	17
Dicamptodon	26	Hydrophidae	23
Dicynodontidae	15	Hyla	30
Diemyctylus	28	Hylidae	10, 30
Dimorphodontidae	12	Hynobiidae	11
Dinosauria	13	Hypsiglena	38
Diplodactylus	50	Ichthyopterygia	15
Diploglossa	46	Ichthyosauridae	15
Dipsosaurus	42	Iguania	47
Discoglossidae	10	Iguanidae	18
Dromicus	35	Iguanodontidae	14
Eastern region	68, 82	Introductory remarks	3
Elapidae	22, 34	Labyrinthodontia	10
Elaps	34	Lacertidae	19, 45
Elasmosauridae	15	Lacertilia	17, 44
Eunydidae	16, 52	Leptoglossa	44
Emys	53	Lichanura	43
Engystoma	30	Lichanuridae	22, 43
Engystomidae	8, 30	Lithodytes	31
Epirhexis	31	Lodia	36
Eumeces	44	Louisianian district	74
Eutaenia	40	Lower Californian region	74, 92
Eretmochelys	51	Macrochelys	52
Faraucia	35	Macrolemmys	53
Faunal regions of the earth	55	Manculus	27
Feyliniidae	20	Megalosauridae	13
Firmisternia	8, 30	Menopoma	25
Floridan district	77	Menopomidae	12, 25
Ganocephala	10	Microsauria	11
Gastrechmia	8	Molgophidae	11
Geeconidae	18, 50	Mosasauridae	20
Geographical distribution in the Regnum Nearcticum, with special reference to the Batrachia and Reptilia	55	Muraenopsis	25
		Najidae	22
		Necturus	24
		Number of species	58
Gerrhonotidae	18, 46	Nyctisaura	18, 50
Gerrhonotus	46	Odontaglossa	9
Goniopholididae	14	Oligosoma	44
Goniopoda	13	Ophibolus	36
Gyalopium	36	Ophidia	33
Gymnophidia	11	Opheosauria	44
Gyrinophilus	28	Opheosaurus	46
Hadrosauridae	14	Ornithosauria	12
Haldea	35	Ornithotarsidae	13
Helicops	43	Orthopoda	13
Heloderma	47	Osceola	36

	Page.
Oudenodontidae	15
Pachyglossa	18
Pacific region	72, 89
Parasuchia	14
Peliontidae	11
Pelodytidae	10
Pelomedusidae	17
Phimothyra	38
Phlegethontiidae	11
Phryniscidae	8
Phrynosoma	49
Phyllodactylus	50
Pipidae	9
Pityophis	39
Placodontidae	14
Plesiosauridae	15
Plethodon	27
Plethodontidae	12, 26
Pleurodelidae	11, 28
Pleurodira	17
Pleurodonta	44
Pleurosternidae	17
Podocnemididae	17
Procoelia	14
Propleuridae	16
Proteida	12, 24
Proteidae	24
Proteroglypha	22, 34
Protorosauridae	15
Protostegidae	16
Pseudemys	52
Pseudobranchus	24
Pterodactylidae	13
Ptyoniidae	11
Pythonidae	21
Pythonomorpha	20
Rana	32
Ranidae	7, 32
Raniformia	7, 32
Relation of distribution to physical causes	93
Relations to other realms	58
Reptilia	12
Rhabdosomidae	22
Rhinenra	44
Rhinochilus	36
Rhinophrynidae	9
Rhiptoglossa	17
Rhynchocephalia	15
Rhynchosauridae	15
Salamandridae	11
Sauromalus	47
Sauropterygia	14

	Page.
Scaphiopidae	10, 31
Scaphiopus	31
Scelidosauridae	13
Sceloporus	48
Scincidae	19, 44
Scolecophidia	21, 44
Sepsidae	19
Sibon	38
Siren	24
Sirenidae	12, 24
Smilisca	31
Solenoglypha	23, 33
Sonora	36
Sonoran region	73, 90
Spea	31
Spelerpes	27
Sphaerodactylus	50
Sphargididae	16, 50
Sphargis	50
Sphenodontidae	15
Spilotes	39
Stegocephali	10
Stenostoma	44
Stenostomidae	21, 44
Stereochilus	27
Sternothaeridae	17
Storeria	42
Symphypoda	13
Tantilla	35
Teidae	19, 45
Teleosauridae	14
Teratosauridae	13
Testudinata	16, 50
Testudinidae	16, 54
Testudo	54
Texan district	80
Thalassochelys	51
The regions of the Regnum nearcticum	67
Thoracosauridae	14
Thoriidae	11
Tortricidae	21
Tortricina	21
Trachystomata	12, 24
Trionychidae	16
Trimorphodon	32
Trogonophidae	20
Tropidoclonium	42
Tropidonotus	42
Tuditanidae	11
Typhlophthalmi	44
Typhlopidae	21
Uma	47

	Page.		Page.
Crodela	11	Works treating of the geographical distribution of North American Batrachia and Reptilia	
Uropeltidae	21		
Uta	48		100
Varanidae	19	Xautusia	45
Verticaria	46	Xenopeltidae	21
Viperidae	23	Xenosauridae	19
Virginia	35	Zonuridae	19
Works on the classification of Batrachia and Reptilia	97		

www.ingramcontent.com/pod-product-compliance
Lightning Source LLC
Chambersburg PA
CBHW031404160426
43196CB00007B/891